CISSP Rapid Review

Darril Gibson

Published with the authorization of Microsoft Corporation by:
O'Reilly Media, Inc.
1005 Gravenstein Highway North
Sebastopol, California 95472

Copyright © 2012 by Darril Gibson
All rights reserved. No part of the contents of this book may be reproduced or transmitted in any form or by any means without the written permission of the publisher.

ISBN: 978-0-7356-6678-8

1 2 3 4 5 6 7 8 9 LSI 7 6 5 4 3 2

Printed and bound in the United States of America.

Microsoft Press books are available through booksellers and distributors worldwide. If you need support related to this book, email Microsoft Press Book Support at mspinput@microsoft.com. Please tell us what you think of this book at *http://www.microsoft.com/learning/booksurvey*.

Microsoft and the trademarks listed at *http://www.microsoft.com/about/legal/en/us/IntellectualProperty/Trademarks/EN-US.aspx* are trademarks of the Microsoft group of companies. All other marks are property of their respective owners.

The example companies, organizations, products, domain names, email addresses, logos, people, places, and events depicted herein are fictitious. No association with any real company, organization, product, domain name, email address, logo, person, place, or event is intended or should be inferred.

This book expresses the author's views and opinions. The information contained in this book is provided without any express, statutory, or implied warranties. Neither the authors, O'Reilly Media, Inc., Microsoft Corporation, nor its resellers, or distributors will be held liable for any damages caused or alleged to be caused either directly or indirectly by this book.

Acquisitions and Development Editor: Kenyon Brown
Production Editor: Rachel Steely
Editorial Production: nSight, Inc.
Technical Reviewer: Andrew Brice
Copyeditor: Richard Carey
Indexer: Nancy Guenther
Cover Design: Best & Company Design
Cover Composition: Zyg Group, LLC
Illustrator: Rebecca Demarest

Contents at a Glance

	Introduction	xxiii
Chapter 1	ACCESS CONTROL	1
Chapter 2	TELECOMMUNICATIONS AND NETWORK SECURITY	25
Chapter 3	INFORMATION SECURITY GOVERNANCE & RISK MANAGEMENT	47
Chapter 4	SOFTWARE DEVELOPMENT SECURITY	89
Chapter 5	CRYPTOGRAPHY	105
Chapter 6	SECURITY ARCHITECTURE & DESIGN	145
Chapter 7	OPERATIONS SECURITY	167
Chapter 8	BUSINESS CONTINUITY & DISASTER RECOVERY PLANNING	195
Chapter 9	LEGAL, REGULATIONS, INVESTIGATIONS, AND COMPLIANCE	217
Chapter 10	PHYSICAL (ENVIRONMENTAL) SECURITY	247
	Index	267

Contents

Introduction xxiii

Chapter 1 Access control — 1

Objective 1.1: Control access by applying the following
concepts/methodologies/techniques...................... 1
- Exam need to know... — 2
- Policies — 2
- Types of controls (preventive, detective, corrective, and so on) — 3
- Techniques (non-discretionary, discretionary, and mandatory) — 5
- Identification and authentication — 6
- Decentralized/distributed access control techniques — 8
- Authorization mechanisms — 9
- Logging and monitoring — 11
- Can you answer these questions? — 12

Objective 1.2: Understand access control attacks 12
- Exam need to know... — 12
- Threat modeling — 13
- Asset valuation — 14
- Vulnerability analysis — 15
- Access aggregation — 15
- Can you answer these questions? — 16

Objective 1.3: Assess effectiveness of access controls.......... 17
- Exam need to know... — 17
- User entitlement — 17
- Access review and audit — 18
- Can you answer these questions? — 19

What do you think of this book? We want to hear from you!

Microsoft is interested in hearing your feedback so we can continually improve our books and learning resources for you. To participate in a brief online survey, please visit:

microsoft.com/learning/booksurvey

Objective 1.4: Identity and access provisioning
lifecycle (e.g., provisioning, review, revocation) **19**

    Exam need to know... **19**

    Provisioning **20**

    Review **21**

    Revocation **22**

    Can you answer these questions? **23**

Answers **23**

    Objective 1.1: Control access by applying the
following concepts/methodologies/techniques **23**

    Objective 1.2: Understand access control attacks **23**

    Objective 1.3: Assess effectiveness of access controls **24**

    Objective 1.4: Identity and access provisioning lifecycle
(e.g., provisioning, review, revocation) **24**

Chapter 2   Telecommunications and network security   25

Objective 2.1: Understand secure network architecture
and design (e.g., IP & non-IP protocols, segmentation) **25**

    Exam need to know... **26**

    OSI and TCP/IP models **26**

    IP networking **29**

    Implications of multilayer protocols **30**

    Can you answer these questions? **31**

Objective 2.2: Securing network components **31**

    Exam need to know... **31**

    Hardware (modems, switches, routers, wireless
access points) **31**

    Transmission media (wired, wireless, fiber) **32**

    Network access control devices (firewalls, proxies) **33**

    End-point security **35**

    Can you answer these questions? **36**

	Objective 2.3: Establish secure communication channels (e.g., VPN, TLS/SSL, VLAN)...............	36
	Exam need to know...	36
	Voice (for example, POTS, PBX, VoIP)	36
	Multimedia collaboration (remote meeting technology, instant messaging)	37
	Remote access (screen scraper, virtual application/desktop, telecommuting)	38
	Data communications	39
	Can you answer these questions?	40
	Objective 2.4: Understand network attacks (e.g., DDoS, spoofing).......................	41
	Exam need to know...	41
	DoS and DDoS	41
	Spoofing	43
	Can you answer these questions?	44
	Answers	44
	Objective 2.1: Understand secure network architecture and design (e.g., IP & non-IP protocols, segmentation)	44
	Objective 2.2: Securing network components	44
	Objective 2.3: Establish secure communication channels (e.g., VPN, TLS/SSL, VLAN)	45
	Objective 2.4: Understand network attacks (e.g., DDoS, spoofing)	45
Chapter 3	**Information security governance & risk management**	**47**
	Objective 3.1: Understand and align security function to goals, mission, and objectives of the organization........	48
	Exam need to know...	48
	Align security function	48
	Can you answer these questions?	49

Objective 3.2: Understand and apply security governance **49**
    Exam need to know... **49**
    Organizational processes (e.g., acquisitions, divestitures, governance committees) **50**
    Security roles and responsibilities **51**
    Legislative and regulatory compliance **52**
    Privacy requirements compliance **52**
    Control frameworks **53**
    Due care **55**
    Due diligence **55**
    Can you answer these questions? **56**

Objective 3.3: Understand and apply concepts of confidentiality, integrity, and availability................... **56**
    Exam need to know... **56**
    Confidentiality **56**
    Integrity **57**
    Availability **58**
    Can you answer these questions? **58**

Objective 3.4: Develop and implement security policy **59**
    Exam need to know... **59**
    Security policies **59**
    Standards/Baselines **60**
    Procedures **61**
    Guidelines **62**
    Documentation **62**
    Can you answer these questions? **63**

Objective 3.5: Manage the information lifecycle (e.g., classification, categorization, and ownership)......... **63**
    Exam need to know... **63**
    Manage the information lifecycle **64**
    Can you answer these questions? **65**

Objective 3.6: Manage third-party governance (e.g., on-site assessment, document exchange and review, process/policy review) **65**
    Exam need to know... **65**

Third-party governance	65
Can you answer these questions?	67

Objective 3.7: Understand and apply risk
management concepts 67

Exam need to know...	67
Identify threats and vulnerabilities	68
Risk assessment/analysis (qualitative, quantitative, hybrid)	69
Risk assignment/acceptance	71
Countermeasure selection	72
Tangible and intangible asset valuation	73
Can you answer these questions?	73

Objective 3.8: Manage personnel security 74

Exam need to know...	74
Employment candidate screening (e.g., reference checks, education verification)	74
Employment agreements and policies	75
Employee termination processes	76
Vendor, consultant, and contractor controls	77
Can you answer these questions?	78

Objective 3.9: Develop and manage security
education, training, and awareness 78

Exam need to know...	78
Security education, training, and awareness	78
Can you answer these questions?	79

Objective 3.10: Manage the security function 79

Exam need to know...	80
Budget	80
Metrics	81
Resources	82
Develop and implement information security strategies	83
Assess the completeness and effectiveness of the security program	83
Can you answer these questions?	84

Answers 84

Objective 3.1: Understand and align security function to goals, mission, and objectives of the organization	85
Objective 3.2: Understand and apply security governance	85
Objective 3.3: Understand and apply concepts of confidentiality, integrity, and availability	85
Objective 3.4: Develop and implement security policy	86
Objective 3.5: Manage the information lifecycle (e.g., classification, categorization, and ownership)	86
Objective 3.6: Manage third-party governance (e.g., on-site assessment, document exchange and review, process/policy review)	86
Objective 3.7: Understand and apply risk management concepts	87
Objective 3.8: Manage personnel security	87
Objective 3.9: Develop and manage security education, training, and awareness	88
Objective 3.10: Manage the Security Function	88

Chapter 4 Software development security   89

Objective 4.1: Understand and apply security in the
software development lifecycle. 89

Exam need to know...	90
Development lifecycle	90
Maturity models	92
Operation and maintenance	93
Change management	93
Can you answer these questions?	94

Objective 4.2: Understand the environment and
security controls 94

Exam need to know...	94
Security of the software environment	95
Security issues of programming languages	98
Security issues in source code (e.g., buffer overflow, escalation of privilege, backdoor)	99
Configuration management	100
Can you answer these questions?	100

Objective 4.3: Assess the effectiveness of software security .. **100**
    Exam need to know... **100**
    Assessment methods **100**
    Can you answer these questions? **102**

Answers **102**
    Objective 4.1: Understand and apply security in the software development lifecycle **102**
    Objective 4.2: Understand the environment and security controls **103**
    Objective 4.3: Assess the effectiveness of software security **103**

Chapter 5 Cryptography   105

Objective 5.1: Understand the application and use of cryptography **105**
    Exam need to know... **106**
    Data at rest (e.g., hard drive) **106**
    Data in transit (e.g., on the wire) **106**
    Can you answer these questions? **107**

Objective 5.2: Understand the cryptographic lifecycle (e.g., cryptographic limitations, algorithm/protocol governance) **107**
    Exam need to know... **108**
    Cryptographic lifecycle **108**
    Algorithm/protocol governance **108**
    Can you answer these questions? **109**

Objective 5.3: Understand encryption concepts **109**
    Exam need to know... **109**
    Foundational concepts **110**
    Symmetric cryptography **111**
    Asymmetric cryptography **113**
    Hybrid cryptography **115**
    Message digests **115**
    Hashing **116**
    Can you answer these questions? **118**

Objective 5.4: Understand key management processes **118**
    Exam need to know... **118**
    Creation/distribution **119**
    Storage/destruction **120**
    Recovery **120**
    Key escrow **120**
    Can you answer these questions? **121**

Objective 5.5: Understand digital signatures **121**
    Exam need to know... **121**
    Purpose and process of digital signatures **121**
    Can you answer these questions? **123**

Objective 5.6: Understand non-repudiation................. **123**
    Exam need to know... **124**
    Non-repudiation **124**
    Can you answer these questions? **124**

Objective 5.7: Understand methods of cryptanalytic attacks .. **124**
    Exam need to know... **125**
    Chosen plaintext **125**
    Social engineering for key discovery **126**
    Brute force (e.g., rainbow tables, specialized/
    scalable architecture) **126**
    Ciphertext only **127**
    Known plaintext **127**
    Frequency analysis **128**
    Chosen ciphertext **128**
    Implementation attacks **128**
    Can you answer these questions? **129**

Objective 5.8: Use cryptography to maintain network security **129**
    Exam need to know... **129**
    Link vs. end-to-end encryption **129**
    SSL and TLS **130**
    IPsec **132**
    Can you answer these questions? **132**

Objective 5.9: Use cryptography to maintain application security **132**
    Exam need to know... **133**

Application security	133
Can you answer these questions?	134

Objective 5.10: Understand Public Key Infrastructure (PKI)....135

Exam need to know...	135
PKI	135
Can you answer these questions?	137

Objective 5.11: Understand certificate related issues 137

Exam need to know...	137
Certificates	137
Validating certificates	138
Can you answer these questions?	139

Objective 5.12: Understand information hiding alternatives (e.g., steganography, watermarking) 139

Exam need to know...	139
Information hiding alternatives	139
Can you answer these questions?	140

Answers 140

Objective 5.1: Understand the application and use of cryptography.	140
Objective 5.2: Understand the cryptographic lifecycle (e.g., cryptographic limitations, algorithm/protocol governance)	141
Objective 5.3: Understand encryption concepts	141
Objective 5.4: Understand key management processes	141
Objective 5.5: Understand digital signatures	142
Objective 5.6: Understand non-repudiation	142
Objective 5.7: Understand methods of cryptanalytic attacks	142
Objective 5.8: Use cryptography to maintain network security	143
Objective 5.9: Use cryptography to maintain application security	143
Objective 5.10: Understand Public Key Infrastructure (PKI)	143
Objective 5.11: Understand certificate related issues	144
Objective 5.12: Understand information hiding alternatives (e.g., steganography, watermarking)	144

Chapter 6   Security architecture & design   145

Objective 6.1: Understand the fundamental
concepts of security models (e.g., Confidentiality,
Integrity, and Multi-level Models)........................145
    Exam need to know...   146
    Security models   146
    Can you answer these questions?   149

Objective 6.2: Understand the components of
information systems security evaluation models 149
    Exam need to know...   149
    Product evaluation models (e.g., Common Criteria)   149
    Industry and international security
    implementation guidelines (e.g., PCI DSS, ISO)   151
    Can you answer these questions?   153

Objective 6.3: Understand security capabilities of
information systems (e.g., memory protection,
virtualization, Trusted Platform Module) 153
    Exam need to know...   153
    Security capabilities of information systems   153
    Can you answer these questions?   155

Objective 6.4: Understand the vulnerabilities of
security architectures 155
    Exam need to know...   155
    System (e.g., covert channels, state attacks, emanations) 155
    Technology and process integration (e.g.,
    single point of failure, service-oriented architecture) 157
    Can you answer these questions?   158

Objective 6.5: Understand software and system
vulnerabilities and threats 158
    Exam need to know...   158
    Web-based (e.g., XML, SAML, OWASP)   158
    Client-based (e.g., applets)   159
    Server-based (e.g., data flow control)   160
    Database security (e.g., inference, aggregation,
    data mining, warehousing)   160
    Distributed systems (e.g., cloud computing,
    grid computing, peer to peer)   162
    Can you answer these questions?   162

Objective 6.6: Understand countermeasure
principles (e.g., defense in depth) **163**
    Exam need to know... **163**
    Defense in depth **163**
    Can you answer these questions? **163**

Answers **164**
    Objective 6.1: Understand the fundamental concepts of security models (e.g., Confidentiality, Integrity, and Multi-level Models) **164**
    Objective 6.2: Understand the components of information systems security evaluation models **164**
    Objective 6.3: Understand security capabilities of information systems (e.g., memory protection, virtualization, trusted platform module) **164**
    Objective 6.4: Understand the vulnerabilities of security architectures **165**
    Objective 6.5: Understand software and system vulnerabilities and threats **165**
    Objective 6.6: Understand countermeasure principles (e.g., defense in depth) **165**

Chapter 7 Operations security **167**

Objective 7.1: Understand security operations concepts **167**
    Exam need to know... **168**
    Need-to-know/least privilege **168**
    Separation of duties and responsibilities **169**
    Monitor special privileges (e.g., operators, administrators) **169**
    Job rotation **170**
    Marking, handling, storing, and destroying of sensitive information **170**
    Record retention **172**
    Can you answer these questions? **172**

Objective 7.2: Employ resource protection **173**
    Exam need to know... **173**
    Media management **173**
    Asset management (e.g., equipment lifecycle, software licensing) **174**
    Can you answer these questions? **174**

Objective 7.3: Manage incident response	**174**
Exam need to know...	**174**
Detection	**175**
Response	**175**
Reporting	**176**
Recovery	**177**
Remediation and review (e.g., root cause analysis)	**177**
Can you answer these questions?	**178**
Objective 7.4: Implement preventative measures against attacks (e.g., malicious code, zero-day exploit, denial of service)	**178**
Exam need to know...	**178**
Attacks	**178**
Preventative measures	**180**
Can you answer these questions?	**182**
Objective 7.5: Implement and support patch and vulnerability management	**182**
Exam need to know...	**182**
Patch management	**182**
Vulnerability management	**183**
Can you answer these questions?	**184**
Objective 7.6: Understand change and configuration management (e.g., versioning, base lining).	**185**
Exam need to know...	**185**
Change management	**185**
Configuration management	**186**
Can you answer these questions?	**187**
Objective 7.7: Understand system resilience and fault tolerance requirements	**187**
Exam need to know...	**188**
Fault tolerance for disks	**188**
Fault tolerance for servers	**190**
Can you answer these questions?	**191**
Answers	**192**
Objective 7.1: Understand security operations concepts	**192**
Objective 7.2: Employ resource protection	**192**
Objective 7.3: Manage incident response	**192**

	Objective 7.4: Implement preventative measures against attacks (e.g., malicious code, zero-day exploit, denial of service)	**193**
	Objective 7.5: Implement and support patch and vulnerability management	**193**
	Objective 7.6: Understand change and configuration management (e.g., versioning, base lining)	**194**
	Objective 7.7: Understand system resilience and fault tolerance requirements	**194**
Chapter 8	**Business continuity & disaster recovery planning**	**195**
	Objective 8.1: Understand business continuity requirements	**195**
	Exam need to know...	**196**
	Develop and document project scope and plan	**196**
	Can you answer these questions?	**198**
	Objective 8.2: Conduct business impact analysis	**198**
	Exam need to know...	**198**
	Identify and prioritize critical business functions	**198**
	Determine maximum tolerable downtime and other criteria	**200**
	Assess exposure to outages (e.g., local, regional, global)	**201**
	Define recovery objectives	**201**
	Can you answer these questions?	**202**
	Objective 8.3: Develop a recovery strategy	**202**
	Exam need to know...	**202**
	Implement a backup storage strategy (e.g., offsite storage, electronic vaulting, tape rotation)	**202**
	Recovery site strategies	**204**
	Can you answer these questions?	**206**
	Objective 8.4: Understand disaster recovery process	**206**
	Exam need to know...	**206**
	Response	**207**
	Personnel	**208**
	Communications	**208**
	Assessment	**209**
	Restoration	**209**
	Provide training	**211**
	Can you answer these questions?	**211**

Objective 8.5: Exercise, assess, and maintain the plan
(e.g., version control, distribution) 211
    Exam need to know...   211
    Exercises   212
    Maintain the plan   213
    Can you answer these questions?   214

Answers 214
    Objective 8.1: Understand business continuity
    requirements   214
    Objective 8.2: Conduct business impact analysis   214
    Objective 8.3: Develop a recovery strategy   214
    Objective 8.4: Understand disaster recovery process   215
    Objective 8.5: Exercise, assess and maintain the
    plan (e.g., version control, distribution)   215

Chapter 9   Legal, regulations, investigations, and compliance   217

Objective 9.1: Understand legal issues that pertain
to information security internationally 217
    Exam need to know...   218
    Computer crime   218
    Licensing and intellectual property (e.g.,
    copyright, trademark)   219
    Import/Export   222
    Trans-border data flow   223
    Privacy   224
    Can you answer these questions?   225

Objective 9.2: Understand professional ethics 225
    Exam need to know...   225
    (ISC)² Code of Professional Ethics   226
    Support organization's code of ethics   227
    Can you answer these questions?   227

Objective 9.3: Understand and support investigations. 228
    Exam need to know...   228
    Policy, roles, and responsibilities (e.g., rules of
    engagement, authorization, scope)   228

Incident handling and response	**230**
Evidence collection and handling (e.g., chain of custody, interviewing)	**230**
Reporting and documenting	**231**
Can you answer these questions?	**232**

Objective 9.4: Understand forensic procedures **233**

Exam need to know...	**233**
Media analysis	**233**
Network analysis	**235**
Software analysis	**236**
Hardware/embedded device analysis	**237**
Can you answer these questions?	**238**

Objective 9.5: Understand compliance requirements and procedures **239**

Exam need to know...	**239**
Regulatory environment	**239**
Audits	**240**
Reporting	**241**
Can you answer these questions?	**241**

Objective 9.6: Ensure security in contractual agreements and procurement processes (e.g., cloud computing, outsourcing, vendor governance) **242**

Exam need to know...	**242**
Contractual agreements and procurement processes	**242**
Can you answer these questions?	**244**

Answers **244**

Objective 9.1: Understand legal issues that pertain to information security internationally	**244**
Objective 9.2: Understand professional ethics	**245**
Objective 9.3: Understand and support investigations	**245**
Objective 9.4: Understand forensic procedures	**245**
Objective 9.5: Understand compliance requirements and procedures	**246**
Objective 9.6: Ensure security in contractual agreements and procurement processes (e.g., cloud computing, outsourcing, vendor governance)	**246**

Chapter 10 Physical (environmental) security   247

Objective 10.1: Understand site and facility design considerations 247
Exam need to know... 248
Site and facility design considerations 248
Can you answer these questions? 249

Objective 10.2: Support the implementation and operation of perimeter security (e.g., physical access control and monitoring, audit trails/access logs) . . . 250
Exam need to know... 250
Physical access control and monitoring 250
Audit trails/access logs 252
Can you answer these questions? 252

Objective 10.3: Support the implementation and operation of internal security (e.g., escort requirements/visitor control, keys and locks) 253
Exam need to know... 253
Escort requirements/visitor control 253
Keys and locks 254
Can you answer these questions? 255

Objective 10.4: Support the implementation and operation of facilities security (e.g., technology convergence) 255
Exam need to know... 255
Communications and server rooms 255
Restricted and work area security 256
Data center security 257
Utilities and heating, ventilation, and air conditioning (HVAC) considerations 257
Water issues (e.g., leakage, flooding) 258
Fire prevention, detection, and suppression 258
Can you answer these questions? 260

Objective 10.5: Support the protection and securing of equipment 260
Exam need to know... 260
Can you answer these questions? 261

Objective 10.6: Understand personnel privacy and
safety (e.g., duress, travel, monitoring) **261**

    Exam need to know... **261**

    Personnel privacy and safety **262**

    Can you answer these questions? **263**

Answers **263**

    Objective 10.1: Understand site and facility
design considerations **263**

    Objective 10.2: Support the implementation
and operation of perimeter security (e.g.,
physical access control and monitoring, audit
trails/access logs) **263**

    Objective 10.3: Support the implementation
and operation of internal security (e.g., escort
requirements/visitor control, keys and locks) **264**

    Objective 10.4: Support the implementation
and operation of facilities security (e.g.,
technology convergence) **264**

    Objective 10.5: Support the protection and
securing of equipment **264**

    Objective 10.6: Understand personnel privacy
and safety (e.g., duress, travel, monitoring) **265**

Index *267*

What do you think of this book? We want to hear from you!

Microsoft is interested in hearing your feedback so we can continually improve our
books and learning resources for you. To participate in a brief online survey, please visit:

microsoft.com/learning/booksurvey

Introduction

This Rapid Review is designed to assist you with studying for the (ISC)² CISSP exam. The Rapid Review series is designed for exam candidates who already have a good grasp of the exam objectives through a combination of experience, skills, and study and could use a concise review guide to help them assess their readiness for the exam.

The CISSP exam is aimed at an IT security professional who has a minimum of five years of direct full-time security work experience in two or more of the 10 domains of the (ISC)² CISSP Common Body of Knowledge (CBK). One year can be waived for certain college degrees and technical certifications.

Candidates who take this exam should have the knowledge and skills required to do the following:

- Identify risk and participate in risk mitigation activities
- Provide infrastructure, application, operational, and information security
- Apply security controls to maintain confidentiality, integrity, and availability
- Identify appropriate technologies and products
- Operate with an awareness of applicable policies, laws, and regulations

It is important to note that real-world experience with security is required prior to earning the CISSP certification and that having practical knowledge is a key component to achieving a passing score.

This book reviews every concept described in the following exam objective domains:

- 1.0 Access Control
- 2.0 Telecommunications and Network Security
- 3.0 Information Security Governance & Risk Management
- 4.0 Software Development Security
- 5.0 Cryptography
- 6.0 Security Architecture & Design
- 7.0 Operations Security
- 8.0 Business Continuity & Disaster Recovery Planning
- 9.0 Legal, Regulations, Investigations and Compliance
- 10.0 Physical (Environmental) Security

This is a Rapid Review and not a comprehensive guide such as the forthcoming *CISSP Training Kit* (Microsoft Press, 2013). The book covers every exam objective on the CISSP exam but will not necessarily cover every exam question. (ISC)² regularly adds new questions to the exam, making it impossible for this (or any) book to provide every answer. Instead, this book is designed to supplement your existing independent study and real-world experience.

If you encounter a topic in this book with which you do not feel completely comfortable, you can visit the links described in the text in addition to researching the topic further by using other websites, as well as consulting support forums.

> **NOTE** The Rapid Review is designed to assess your readiness for the CISSP exam. It is not designed as a comprehensive exam preparation guide. If you need that level of training for any or all of the exam objectives covered in this book, we suggest the forthcoming *CISSP Training Kit* (ISBN: 9780735657823). The Training Kit will provide comprehensive coverage of each CISSP exam objective, along with exercises, review questions, and practice tests. The Training Kit will also include a discount voucher for the exam.

(ISC)² professional certification program

(ISC)² professional certifications cover the technical skills and knowledge needed to succeed in different IT careers. The CISSP certification is a vendor-neutral credential. An exam is an internationally recognized validation of skills and knowledge and is used by organizations and professionals around the globe. (ISC)² CISSP certification is ISO 17024 Accredited (Personnel Certification Accreditation) and, as such, undergoes regular reviews and updates to the exam objectives. (ISC)² exam objectives reflect the subject areas in an edition of an exam and result from subject matter expert workshops and industry-wide survey results regarding the skills and knowledge required of a professional with a number of years of experience.

> **MORE INFO** For a full list of (ISC)² certifications, go to *https://www.isc2.org/credentials/*.

Acknowledgments

Books like this are never by just one or two people but instead are created with the combined efforts of a large group of people. I'm grateful for the help and support I received from multiple individuals at O'Reilly and Microsoft Press. I'm especially grateful for the outstanding technical input provided by Andrew Brice, the technical editor.

Support & feedback

The following sections provide information on errata, book support, feedback, and contact information.

Errata

We've made every effort to ensure the accuracy of this book and its companion content. Any errors that have been reported since this book was published are listed on our Microsoft Press site at oreilly.com: *http://go.microsoft.com/FWLink/?Linkid=272758*

If you find an error that is not already listed, you can report it to us through the same page.

If you need additional support, email Microsoft Press Book Support at *mspinput@microsoft.com*.

Please note that product support for Microsoft software is not offered through the addresses above.

We want to hear from you

At Microsoft Press, your satisfaction is our top priority and your feedback our most valuable asset. Please tell us what you think of this book at *http://www.microsoft.com/learning/booksurvey*.

The survey is short, and we read every one of your comments and ideas. Thanks in advance for your input!

Stay in touch

Let's keep the conversation going! We're on Twitter: *http://twitter.com/MicrosoftPress*.

Preparing for the Exam

Certification exams are a great way to build your resume and let the world know about your level of expertise. Certification exams validate your on-the-job experience and product knowledge. Although there is no substitute for on-the-job experience, preparation through study and hands-on practice can help you prepare for the exam. We recommend that you augment your exam preparation plan by using a combination of available study materials and courses. For example, you might use the Rapid Review and another training kit for your "at home" preparation and take an (ISC)[2] CISSP professional certification course for the classroom experience. Choose the combination that you think works best for you.

CHAPTER 1

Access control

The Access Control domain covers a variety of different controls used to identify subjects, authenticate them, and control the access they are granted to different objects by controlling rights and permissions. Audit trails are an important element of accounting and logging and, combined with effective authentication, provide individual accountability. Access control attacks are common, and it's important for security professionals to have a basic understanding of evaluating threats and analyzing vulnerabilities to determine overall risk. Ideally, access controls are implemented to fully support an organization's security policy, and a way to verify this is through access reviews and audits. These reviews and audits can also detect problems in the identity and access provisioning life cycle, such as inactive accounts that have not been disabled.

This chapter covers the following objectives:

- Objective 1.1: Control access by applying the following concepts/methodologies/techniques
- Objective 1.2: Understand access control attacks
- Objective 1.3: Assess effectiveness of access controls
- Objective 1.4: Identity and access provisioning lifecycle (e.g., provisioning, review, revocation)

Objective 1.1: Control access by applying the following concepts/methodologies/techniques

For this exam objective, you must understand many of the basics related to IT risk management. Security policies provide overall direction for an organization. Personnel within the organization then implement different types of controls to support the policy. At the core of access control is effective authentication, and it's important to understand the authentication factors. Without effective authentication, it isn't possible to enforce authorization and data within audit trails is not useful. Single sign-on (SSO) authentication methods have been widely available within a single organization's environment, but newer methods support SSO in federations. You should have an understanding of federated identity management systems and some of the XML-based protocols that they use to share authentication information.

Exam need to know...

- **Policies**
  For example: Do you know and understand the common elements of a security policy, such as the principle of least privilege and separation of duties? What is an acceptable use policy?
- **Types of controls (preventive, detective, corrective, and so on)**
  For example: What type of control is an audit trail? What type of control is a security guard?
- **Techniques (for example, non-discretionary, discretionary, and mandatory)**
  For example: What is a commonly used non-discretionary model that organizes users into groups?
- **Identification and authentication**
  For example: One authentication method requires users to enter a password and a PIN. Another model requires users to use a smart card and a PIN. Which is stronger?
- **Decentralized/distributed access control techniques**
  For example: What XML-based standards are used to provide SSO capabilities in a federated identity management system?
- **Authorization mechanisms**
  For example: How does a constrained user interface control access? What are the differences among an access control list, a capability table, and an access control matrix?
- **Logging and monitoring**
  For example: What is provided by an audit trail? What is included in log management?

Policies

At the heart of any access control strategy is one or more security policies that identify the overall security goals of an organization. The security policy provides a high-level overview of generalized goals, and it is used to create more specific guidelines, standards, and procedures. Security policies commonly include one or more of the following elements:

- An acceptable use policy (AUP) informs users of their responsibilities when using IT systems and identifies unacceptable behaviors. Users should reread and acknowledge the AUP periodically, such as once a year.
- Least privilege refers to the practice of granting subjects access only to what they need to perform their jobs and no more.
- A separation of duties policy ensures that no single entity can control an entire process. It helps prevent fraud by requiring two or more people to conspire together.
- Job rotation policies help prevent fraud by ensuring that a person does not remain in the same position for an extended period and gain excessive control over any area of the business.

Access Control policies commonly refer to subjects and objects. A subject (such as a user) can access an object (such as a file or other resource). Subjects are often grouped together by using roles or groups to simplify administration. Similarly, objects are also grouped together, such as grouping files within folders or shares to simplify administration. As a best practice, permissions for an object are rarely granted to a single subject.

True or false? The primary goal of security policies is to protect confidentiality, integrity, and availability of an organization's assets.

Answer: *True*. Security policies and procedures are in place to support the core security goals of preventing the loss of confidentiality, integrity, or availability of assets.

> **EXAM TIP** Often a question can sound overly complex when it is referring to *subjects* and *objects*, but you can usually simplify it by substituting the word *users* for *subjects* and the word *files* for *objects*. For example, instead of "Access Control administration is simplified by grouping subjects and objects," you can think of this as "Access Control administration is simplified by grouping users and files." Subjects can be more than just users, and objects can be more than just files, but substituting subjects with users and objects with files works in many instances without changing the meaning.

> **MORE INFO** The Operations Security domain (covered in Chapter 7, "Operations security") specifically mentions some security operations concepts, such as need-to-know, least privilege, separation of duties, and job rotation. Overall, you'll find that many of the Access Control topics have some similarities to the Operations Security topics.

Types of controls (preventive, detective, corrective, and so on)

Security controls are safeguards or countermeasures put into place to reduce overall risk. One way they are classified is based on how they are implemented:

- **Technical or logical** controls are implemented with technology such as protecting objects with permissions or requiring users to change their passwords with a technical password policy.
- **Physical** controls include elements that you can physically touch, such as a door lock or a closed circuit television (CCTV).
- **Administrative or management** controls are written security policies or methods used to check the effectiveness of security, such as assessment or audit.

True or false? Pre-employment background investigations are a type of administrative control.

Answer: *True*. This is a procedure and would be done based on a policy.

Another way controls are classified is based on what they do, and they are often grouped together with the implementation method. Control classifications include the following:

- **Preventive** controls attempt to prevent incidents before they occur. A firewall is a technical preventive control because it can prevent malicious traffic from entering a network. A guard is a physical preventive control. Administrative preventive controls include access reviews and audits.
- **Detective** controls identify security violations after they have occurred, or they provide information about the violation as part of an investigation. An intrusion detection system is a technical detective control, and a motion detector is a physical detective control. Note that both an intrusion detection system and a motion detector include the word "detect," which is a good clue. Reviewing logs or an audit trail after an incident is an administrative detective control.
- **Corrective** controls attempt to modify the environment after an incident to return it to normal. Antivirus software that quarantines a virus is an example of a technical corrective control. A fire extinguisher is an example of a physical corrective control.
- **Deterrent** controls attempt to discourage someone from taking a specific action. A high fence with lights at night is a physical deterrent control. A strict security policy stating severe consequences for employees if it is violated is an example of an administrative deterrent control. A proxy server that redirects a user to a warning page when a user attempts to access a restricted site is an example of a technical deterrent control.
- **Directive** controls are administrative controls that provide direction or guidance.
- **Compensating** controls are controls used as alternatives to the recommended controls. NIST SP800-53 mentions a compensating control used for an industrial control system (ICS). A change management policy might dictate the testing of all updates on live systems prior to deployment, but this might not be feasible for an ICS. A compensating control is an offline replicated system used for testing.
- **Recovery** controls provide methods to recover from an incident.

True or false? A user entitlement access review and audit is a detective control.

Answer: *False*. It is a preventive control. It is designed to identify whether users have more privileges than necessary prior to an incident. Discrepancies in assigned privileges can be corrected to prevent an incident. If an incident had already occurred, reviewing an audit trail would be a detective control.

> **EXAM TIP** Given a security control, you should be able to identify it as preventive, detective, corrective, deterrent, recovery, or directive in nature. For example, visible security controls are deterrent in nature because they deter attackers. You should also be able to identify it as technical, physical, or administrative. For example, motion-activated lights are a physical preventive control.

> **MORE INFO** National Institute of Standards and Technology (NIST) Special Publication (SP) 800-53, "Information Security," provides in-depth coverage of security controls. You can download it from this page: *http://csrc.nist.gov/publications/PubsSPs.html*. NIST SP 800-53 organizes the controls in families and classes that don't directly relate to the CISSP types of preventive, detective, and corrective, but the document does provide excellent descriptions and information about usage of IT security controls.

Techniques (non-discretionary, discretionary, and mandatory)

Non-discretionary access controls are centrally managed, and discretionary access controls (DAC) are managed by data owners. Mandatory access controls (MACs) are predefined by a higher authority, such as a policy that defines access labels.

In a DAC model, every object is owned by a subject and the owner has full control over the object. For example, when a user creates a file, the user owns the file and can modify the permissions. Common operating systems such as Windows and Linux use the DAC model.

In non-DAC models, subject and object access is controlled centrally, such as by an administrator. Role-Based Access Control (RBAC) is a common example in which subjects are placed into roles or groups by administrators. Access to objects is granted to the roles rather than to individuals.

True or false? Access control administration is simplified by grouping subjects and grouping objects.

Answer: *True*. Users and other subjects are often grouped together by using an RBAC model. Similarly, objects such as files are often grouped together in folders and shares.

> **EXAM TIP** Know the primary differences between the models. DAC grants full control to end users. MAC uses predefined rules and labels to grant access. RBAC is a non-discretionary model that is controlled by administrators.

MAC uses labels assigned to subjects and objects, and when the labels match, subjects are granted access. Labels can be assigned in a hierarchical environment such as Unclassified, Secret, and Top Secret, with higher-level authorization also providing access to lower-level classifications. For example, someone granted Top Secret access also has Secret access. Labels can be assigned in a compartmentalized environment where access to one compartment does not provide access to any other data. A hybrid model uses compartments within classification levels and is easier to manage.

True or false? The Bell-LaPadula model is an example of a MAC model.

Answer: *True*. It uses a basic rule of no read up, no write down. The Simple Security Rule states that a subject cannot read up, and the *-property (star property) rule states that a subject cannot write down. In contrast, the Biba model (which is also a MAC model) uses a Simple Integrity Axiom of no read down and a * (star) Integrity Axiom of no write up.

EXAM TIP Different access control models have different primary goals. For example, the Bell-LaPadula model has a primary focus of ensuring confidentiality (ensuring that unauthorized users cannot access data). The Biba model has a primary focus on integrity (ensuring that unauthorized modification of data does not occur). The Clark-Wilson model and the Chinese Wall (also called Brewer-Nash) model use strict separation of duties rules to enforce integrity.

Identification and authentication

Identification occurs when a user claims an identity, such as with a user name. Authentication occurs when the user proves the claimed identity by using one or more factors of authentication. The three primary factors of authentication are as follows:

- Something you know (such as a password or PIN)
- Something you have (such as a smart card or RSA token)
- Something you are (proven with biometrics)

You can combine two or more factors to provide stronger authentication. Two-factor authentication uses a method in two of the categories and is stronger than using a single factor. Multifactor authentication uses methods in two or more categories.

Even though passwords are usually stored as a hash, they can be cracked by using common comparative analysis tools. If attackers can access a password database, they can perform an offline analysis and quickly crack the passwords. Rainbow tables are commonly used in these attacks. Salting the hash with random bits protects against many offline password attacks, including the use of rainbow tables.

MORE INFO A hashing algorithm provides one-way encryption of data, creating a unique fixed-size string. Strong hashing algorithms such as SHA-256 are resistant to collisions, meaning that it is not feasible to identify another input to create the same hash. In this case, it would be difficult to identify another password that creates the same hash so that a successful comparative analysis discovers the original password. Salts thwart rainbow table attacks, but other attacks, such as dictionary attacks or brute-force attacks, can still be successful against an offline database of passwords.

True or false? Using a PIN and a password is an example of multifactor authentication.

Answer: *False*. A personal identification number (PIN) and a password are both in the same authentication factor (something you know). This is one-factor authentication, not multifactor authentication.

EXAM TIP Know the differences between the authentication factors. Passwords are the weakest form of authentication, but combining passwords with a method in a different factor, such as a smart card or using biometrics, provides multifactor authentication and increases the strength. Some specific access methods require stronger authentication. For example, remote access connections are vulnerable to attack, and multifactor authentication makes it more difficult for an attacker to impersonate a user.

Biometrics provide strong authentication, but they are susceptible to both false positives and false negatives. A false positive presents the highest risk. This occurs when an unauthorized individual is incorrectly identified as being authorized. The accuracy of a biometric system is identified by the crossover error rate (CER), which is calculated from Type 1 errors (false rejections) and Type 2 errors (false positives). A lower CER indicates a more accurate biometric system.

A simple way to reduce the risk of Type 1 and Type 2 errors is to use two-factor authentication. For example, in addition to the biometric method, the user can also be required to use a password.

A similar concept is used with credit cards and online purchases. Instead of just requiring the user to provide the credit card number and expiration date, users are often required to provide the credit card verification code. This is a 3-digit or 4-digit number on the front or back of the card.

True or false? Between iris scanners and retinal scanners, iris scanners are the most accurate form of biometric authentication.

Answer: *False*. Retinal scans are the most accurate form of biometric authentication. Even identical twins will have identifiable differences.

> **EXAM TIP** Biometric authentication is the strongest form of authentication. Retinal scans measure the blood vessels in the back of the eye and are the most accurate method. Fingerprints are the most common method of biometrics in use.

Single sign-on (SSO) techniques are used in several different access control and identity management systems. These allow a user to log on once and access multiple resources without logging on again.

Internal networks can use a database such as Microsoft's Active Directory to manage user identities and provide SSO. Regular users have a single account and can access any resources in the network as long as they have permissions.

Kerberos is commonly used as an authentication protocol in a centralized model. It requires a central database of accounts and synchronized time (ideally synchronized with an external time source). Kerberos uses time-stamped tickets to authenticate accounts when they try to access a resource. These tickets are encrypted with symmetric encryption. Early versions of Kerberos used Data Encryption Standard (DES), which is now considered cracked, and current versions use Advanced Encryption Standard (AES).

> **EXAM TIP** Know that SSO is part of an identity management system and can be used in a centralized environment such as a Microsoft domain using Kerberos. SSO methods are also implemented in federated identity management systems that include different operating systems.

Remote access protocols provide authentication, authorization, and accounting (AAA) services. Some common AAA protocols include the following:

- **Terminal Access Controller Access-Control System (TACACS)**   One of the first AAA protocols used with remote access systems, TACACS has been replaced by RADIUS, TACACS+, or Diameter in most situations. TACACS uses UDP port 49 by default.

- **Remote Authentication Dial-in User Service (RADIUS)**   This is a widely used AAA protocol in remote access systems and by Internet service providers (ISPs). It is used with both dial-in and virtual private network (VPN) access. It uses UDP and encrypts the password but not the entire authentication session.

- **TACACS Plus (TACACS+)**   Cisco created this as a proprietary upgrade to TACACS. It separates each element of AAA in three processes. In comparison, RADIUS combines authentication and authorization. TACACS+ uses TCP port 49 instead of UDP, and it encrypts the entire authentication session instead of just the password.

- **Diameter**   This is an alternative or upgrade to RADIUS, and it has much more flexibility. It can be used with wireless devices, Voice over IP (VoIP), Mobile IP, and smartphones, but it is not backward-compatible with RADIUS. The name implies it is twice as good as RADIUS because the diameter of a circle is twice the length of the radius.

EXAM TIP   Common remote access protocols are RADIUS, TACACS+, and Diameter. TACACS+ is proprietary to Cisco and includes several benefits over RADIUS. Diameter has the most flexibility.

MORE INFO   RADIUS is defined in RFC 2865 and has been updated by 2868, 3575, and 5080. Diameter is defined in RFC 3588. By replacing *xxxx* with the RFC number, you can view any RFC with the following URL: *http://tools.ietf.org/html/rfcxxxx*. TACACS+ is not defined in a formal RFC, but it is documented in a draft document available here: *http://tools.ietf.org/html/draft-grant-tacacs-02*.

Decentralized/distributed access control techniques

Distributed computing environments (DCEs) use distributed SSO mechanisms to control access. Federated identity management systems are used to provide SSO to Internet users from different entities. In this context, a federation is a group of companies that decide they want to collaborate to share resources.

For example, imagine that employees in Company A are granted access to resources in Company B and Company C. Instead of requiring these users to have three separate passwords, they can log on once within Company A and then access resources in Company B and Company C without logging on again.

MORE INFO   The MSDN article "Federated Identity: Scenarios, Architecture, and Implementation" includes excellent examples of when a federated identity management system is needed and of some of the challenges. You can access it here: *http://msdn.microsoft.com/library/aa479079.aspx*.

A significant challenge is sharing the authentication and authorization information between the companies. If everyone used the same technologies, it would be easier to share the data, but more often, the federation has a heterogeneous identity environment. Different companies use different identity management methods.

Standards based on Extensible Markup Language (XML) are often used to share federated identity information over the Internet. Some of the commonly used standards include the following:

- Security Assertion Markup Language (SAML), which includes both authentication and authorization information
- Service Provisioning Markup Language (SPML), which is used to share provisioning information between organizations in the federation
- Extensible Access Control Markup Language (XACML), which provides a standard for evaluating authorization requests

True or false? SAML is used to provide SSO access when users are accessing sites with web browsers.

Answer: *True*. SAML is one of the schemas used with federated access, and it is often used to provide access via web browsers.

> **EXAM TIP** A federated identity is an SSO-based identity that is portable between different organizations within a federation. Federated identity management systems use HTML-based and XML-based languages to share authentication and authorization information with each other. Many e-commerce solutions have implemented federated identity management systems to streamline the customer experience when making purchases from partner companies in a federation.

> **MORE INFO** There are several resources on SAML, SPML, and XACML that you might find valuable. "Demystifying SAML" is available here: *http://www.oracle.com/technetwork/articles/entarch/saml-084342.html*. "XACML Overview" is available here: *http://xml.coverpages.org/xacml.html*. The full SPML v2.0 specification can be downloaded here: *http://www.oasis-open.org/committees/download.php/17708/pstc-spml-2.0-os.zip*.

Authorization mechanisms

After subjects prove their identity, or authenticate, they are granted access to objects based on their proven identity. Most authorization systems start with an implicit deny philosophy. For example, a user is denied access to files and folders unless the user is specifically given permissions to access them.

A common type of access model is RBAC, in which users are placed into roles or groups and privileges are granted to the group. However, there are other authorization mechanisms that can be used either separately or in combination with an RBAC model.

A constrained user interface limits what the user can see or do based on the user's privileges. For example, imagine an application that can be used by both administrators and regular users. When administrators use it, all the menu items are visible and the application has full functionality. When a regular user starts the application, the menu items are either hidden or dimmed so that they can't be selected. When menu items are hidden, the users are unaware of the advanced capabilities.

Databases commonly use views as a constrained user interface to limit the available data. For example, an employee table might include names, addresses, phone numbers, and salary data. One view might include only names and phone numbers, and another view might include names and salary data. Users are granted access to the view that shows data that they're authorized to view, but they are not granted access to the other view or the underlying table.

Temporal-based authorization controls limit access based on time. For example, a virtual private network (VPN) user might be authorized to connect any weekday between 7:00 A.M. and 7:00 P.M. If the user attempts to connect on a weekend, the connection is blocked.

Location-based authorization controls limit access to specific locations. For example, an employee might be authorized to work from home by using a dial-in connection. Caller ID or callback technologies can be used to ensure that the user is calling from home and that another user is not impersonating the user from another location.

An extension of this is location-based authorization controls using domains. For example, the United States government purchased antivirus (AV) software for all government employees that they can download for free. The AV vendors restrict access to the download websites to allow only traffic coming from .mil domain locations.

Access control lists (ACLs), capability tables, and access control matrices (ACM) are related. An ACL is directly associated with an object, a capability table is directly associated with a subject, and an ACM combines them both. For example, a folder named data (an object) includes an ACL that lists all users granted access to the folder and their specific levels of access, such as read or write. A capability table might be created for a user named Darril (the subject) and include a list of all folders that he can access. The ACM includes all objects and all subjects and can be quite large.

NOTE ACLs are commonly associated with routers as a list of IP addresses, ports, and protocol IDs that are allowed in or out of a network. In this context, the router is the object and the IP address, port, or protocol ID is the subject. An ACL is created for each router, defining the access for different subjects (IP addresses, ports, or protocol IDs). A capability model might be created for different protocols, such as File Transfer Protocol (FTP), and list all routers that allow FTP traffic. An ACM would include all routers and all protocols.

True or false? The security kernel of an operating system controls access between subjects and objects.

Answer: *True*. A security kernel controls access. For example, the Windows kernel-mode security reference monitor in current Windows–based systems uses discretionary ACLs (DACLs) to determine access with the DAC model.

> **EXAM TIP** Know the differences among an ACL, a capability table, and an ACM. The ACL is focused on an object and lists subjects that can access it. The capability table is focused on a subject and lists objects the subject can access. The ACM combines them both.

Logging and monitoring

Logs record activity as it occurs and record details such as who did it (which account), what happened, when it happened, and where it happened. You can use one or more audit logs to create an audit trail. An audit trail is a detective control and provides enough information so that you can identify the relevant events leading up to and during an incident.

Audit trails are required to ensure accountability and depend on effective identification and authorization techniques. If users can easily use another account, the audit trail cannot effectively identify who took an action.

Log management methods ensure that logs are maintained to provide a full audit trail, that the logs are protected from modification, and that they are regularly reviewed. Protecting logs from modification is especially important if they will be used as evidence in court. Access to the logs should be restricted to administrators and security personnel only.

Administrators also use logs to manage and maintain systems. They provide key information used during troubleshooting and recovery of systems after a failure. These logs are used to help prevent or minimize loss of availability.

True or false? Audit trails are a type of preventive control that record who took an action, what action the user took, and when the user took it.

Answer: *False*. Audit trails are a type of detective control. An audit trail logs events as they occur, including details on who, what, when, and where. After an incident has occurred, these logs can be examined to re-create the events.

> **EXAM TIP** Log management includes the practice of reviewing the logs (either manually or with an automated system) and protecting the logs from modification after they've been created. Logs should be synchronized with an external time source so that all logs record the same time. They should be stored in a central location that has restricted access. Attackers will often try to erase their recorded activity, but this is more difficult if the log is stored in another location.

Intrusion detection systems (IDSs) and intrusion prevention systems (IPSs) are also used for logging and monitoring. An IDS is a detective control that can detect attacks, and an IPS is a preventive control that can prevent attacks by detecting and blocking them before they reach an internal network. Both controls send alerts or some type of notification when they detect a potential attack.

However, each alert isn't necessarily an attack. IDSs have adjustable thresholds, and an alert is created only when activity exceeds the threshold. If the threshold is too high, actual attacks can get through undetected. If the threshold is too low, the system generates too many false positives.

True or false? An IPS is placed in line with traffic.

Answer: *True*. All traffic goes through an IPS. The IPS detects and blocks malicious traffic but allows safe traffic through to the network.

> **MORE INFO** IDSs and IPSs are mentioned in the Operations Security domain in Chapter 7, including methods of detection and response.

Can you answer these questions?

You can find the answers to these questions at the end of the chapter.

1. An organization has created a high-level document designed to provide direction to employees about security within the organization. What is this?
2. An audit trail is being used to identify events leading up to a security incident. What type of control is an audit trail in this situation?
3. What is the difference between an ACL and an ACM?
4. What is a measure of a secure biometric authentication system?
5. What is the purpose of SAML?
6. What type of authorization mechanism is a database view?
7. An audit trail is used after an incident. What is required for this audit trail to support individual accountability?

Objective 1.2: Understand access control attacks

Risk is identified by calculating the probability that a threat will exploit a vulnerability. Often, the threats come in the form of attackers attempting to exploit vulnerabilities in an organization's people, processes, or technology. Risk management includes identifying threats by using threat modeling, identifying valuable assets to protect, and analyzing vulnerabilities. Risks can then be mitigated with controls that reduce the impact of threats or reduce vulnerabilities.

Exam need to know...

- Threat modeling
  For example: What are some methods of social engineering? What's the difference between a denial of service (DoS) and a distributed denial of service (DDoS) attack?

- Asset valuation
  For example: When should the value of assets be identified? What assets should be evaluated within an organization as part of a risk management process?
- Vulnerability analysis
  For example: How often should vulnerability assessments be done? What can a vulnerability scan detect?
- Access aggregation
  For example: What types of attacks are launched by Advanced Persistent Threats (APTs)? Who can be a target of an APT?

Threat modeling

Threat modeling is the process of identifying potential and realistic threats to an organization's assets. You should be aware of common methods of access control attacks, including the following:

- **Social engineering** Attackers can often gain access simply by asking. This includes in-person, over the phone, and via email such as with phishing, spear-phishing, and whaling. It can also include tailgating and shoulder surfing.
- **Dumpster diving** If papers are thrown in the trash, they can easily be retrieved to gain information.
- **Malware** Viruses, worms, Trojan horses, and logic bombs are common methods that attackers use to gain control of a system or launch access control attacks.
- **Mobile code** Attackers have hijacked legitimate websites and installed malicious ActiveX and Java scripts. This represents a threat to the organization hosting the website. Additionally, visitors can be attacked by a drive-by download.
- **Denial of Service (DoS)** These come from a single attacker and attempt to disrupt normal operation or service of a system. A classic DoS attack is the SYN flood attack. DoS attacks are commonly launched against Internet-facing servers (any server that can be reached by another public IP address).
- **Distributed DoS (DDoS)** These come from multiple attackers, such as zombies in a botnet.
- **Buffer overflow** When input validation isn't used, unexpected code can cause an unhandled error and allow an attacker to install malicious code on a system.
- **Password crackers** Applications are widely available that can crack a password through comparative analysis. If the attacker can gain access to a database with passwords, the attacker can crack the passwords offline.
- **Spoofing** Attackers attempt to impersonate others in many different ways. They can spoof IP addresses, MAC addresses, and email addresses. Similarly, masquerading is when a social engineer impersonates someone such as a repairman.

- **Sniffers** Protocol analyzers placed on a network can capture traffic for later analysis. If passwords or valuable data are sent unencrypted, they can easily be read. Sniffers are often used in man-in-the-middle and replay attacks.
- **DNS-related attacks** Users can be tricked into providing their credentials on a bogus website after a DNS poisoning attack redirects traffic. DNS poisoning is used in pharming attacks.

True or false? Executives can be targeted through a whaling attack.

Answer: *True*. Whaling is a form of phishing that targets executives such as CEOs, presidents, and vice presidents.

True or false? A SYN flood attack uses spoofed IP addresses and causes a buffer overflow.

Answer: *False*. A SYN flood attack commonly uses spoofed IP addresses, but it doesn't cause a buffer overflow. Instead, it disrupts the three-way TCP handshake process by holding back the third packet.

> **EXAM TIP** Know the common methods of attacks and the methods used to mitigate them. For example, the best prevention against social engineering is education, malware is detected and isolated with up to date antivirus software, and many DoS and DDoS attacks can be detected or prevented with IDSs or IPSs.

> **MORE INFO** Attacks come from multiple sources. In "Rethinking the Cyber Threat - A Framework and Path Forward," Scott Charney of the Microsoft Trustworthy Computing Group outlines four categories: conventional cybercrimes, in which attackers target systems for criminal purposes; military espionage, in which nation states sponsor attacks against military targets to obtain military secrets; economic espionage, in which intellectual property is stolen for economic gain; and cyber warfare, in which attackers attempt to disrupt or disable the IT services of an enemy. You can download the white paper here: *http://www.microsoft.com/download/details.aspx?id=747*.

Asset valuation

One of the first steps in risk management is identifying the value of assets within the organization. This includes hardware assets, software assets, data and information assets, system assets, and personnel assets.

Key steps within the risk management process depend on knowing the value of the assets. For example, a cost-benefit analysis helps determine the return on investment (ROI) of a control. The ROI is high if you purchase an effective control for US$1,000 to protect a web farm generating 1 million dollars a day. It is ridiculously low if you pay US$1,000 to protect a US$15 keyboard. These examples represent two extremes where the answer is obvious, but the answers aren't always so clear, especially if the value of assets is not known.

True or false? Asset valuation is done only on hardware assets.

Answer: *False*. Asset valuation should be done on all assets, including hardware, software, data or information, and personnel. Many systems, such as a web farm, include hardware, software, and data and represent a combined value much greater than that of their individual components.

> **EXAM TIP** Identifying the value of assets is an important first step in risk management and requires input from management. Technicians managing systems don't necessarily realize how much money a system generates, and without guidance, they might give the same amount of attention to a simple file server as they do to a server generating millions in revenue.

Vulnerability analysis

A vulnerability analysis helps determine how vulnerable a system is to one or more threats. This is often referred to as two separate processes: vulnerability scans and vulnerability assessments.

Vulnerability scans are performed with automated tools such as Nmap to determine what vulnerabilities exist at any given time. Vulnerability scanners can detect a wide assortment of vulnerabilities, including open ports, unpatched or misconfigured systems, and weak passwords.

A vulnerability assessment is an overall examination of the organization beyond just a technical scan. It will often attempt to match threats with vulnerabilities and use available data to determine the likelihood or probability that a threat will attempt to exploit a vulnerability. Data reviewed in an assessment includes security policies, historical data on past incidents, audit trails, and the results of various tests, including vulnerability scans.

Threats and the environment regularly change, so these reviews and scans must be repeated. Based on their security policies and available resources, organizations must decide how often to repeat the vulnerability scans and assessments. For example, a large organization might perform vulnerability scans weekly and vulnerability assessments annually, but a smaller organization might do scans only monthly.

True or false? Risk management is an ongoing process, and a vulnerability analysis is a point-in-time assessment.

Answer: *True*. Risk management is a continuous process that needs regular attention. A vulnerability analysis identifies vulnerabilities at a given time, but changes in threats or the environment negate the findings.

> **EXAM TIP** Controls are put into place to support an organization's security policies. Vulnerability assessments and scans evaluate the effectiveness of these controls.

Access aggregation

Access aggregation refers to the combination of methods used to gain progressively more and more access. As a basic example, malware often attempts to progressively increase its privileges until it has full administrative access. On a larger scale, attackers often use a combination of methods to gain more and more access to an organization.

For example, an attacker might decide to target an organization and start with a dozen or so social engineering phone calls. Each call gets one more piece of information, and eventually the attacker has the names and email addresses of several executives. He might then use whaling to send one or more malware-infected phishing emails to these executives. If one of the executives takes the bait, the malware begins collecting information and sending it to the attacker.

This is challenging enough if you are considering only one attacker. Advanced Persistent Threats (APTs) are composed of full teams of attackers. They often have unlimited funding from a nation-state sponsor, but they could just as easily be funded by any group that has the money and a target.

True or false? An APT is a group of attackers, often sponsored by a government, that attacks only military or government targets.

Answer: *False.* An APT is often sponsored by a government, but it can target any organization. Attacks against organizations such as Google and Lockheed Martin are believed to have come from APTs.

> **EXAM TIP** Attacks previously were primarily opportunistic, where the attackers looked for easy targets. When they failed to breach a target, they moved on to an easier one. In contrast, APTs launch targeted attacks against specific organizations and are not deterred by initial failures. They are often composed of state-sponsored personnel and use a wide variety of attack methods to progressively increase their access.

> **MORE INFO** An article posted in the SANS Institute InfoSec Reading Room, "A Detailed Analysis of an Advanced Persistent Threat Malware" (*http://www.sans.org/reading_room/whitepapers/malicious/detailed-analysis-advanced-persistent-threat-malware_33814*), provides insight into the multiple techniques used by APTs. Malware delivered in a targeted spear-phishing email to a political figure was undetectable as malicious by 29 out of 44 antivirus engines. When successfully installed, it made significant system modifications, installed three Trojan spies, and began capturing and sending data to a command-and-control server. Of course, this would be only the beginning. Information gathered from the attack would be used to launch additional attacks by the APT.

Can you answer these questions?

You can find the answers to these questions at the end of the chapter.

1. An attacker is able to enter data into a webpage and install malware on the system. What should have been done to prevent this?
2. What assets should be evaluated when identifying asset values?
3. What is the primary purpose of vulnerability scans?
4. Who can be a target of an APT?

Objective 1.3: Assess effectiveness of access controls

Access controls should limit access to resources to only the people who need those resources. Two important elements of assessing the effectiveness of the controls are examining user entitlement and performing periodic access reviews and audits.

Exam need to know...

- User entitlement
  For example: Which accounts deserve the most attention when considering user entitlement?
- Access review and audit
  For example: How can you verify whether the principle of least privilege is being enforced?

User entitlement

User entitlement refers to the privileges granted to users when their accounts are first created and during the lifetime of the accounts. One of the primary considerations is ensuring that the principle of least privilege is followed. Users should not have access to more privileges than they need to perform their jobs.

Managing changes during the lifetime of the account can be challenging. Often, the process requires users to submit a request that must go through an approval process, and during this time, the user isn't able to complete job requirements. Bypassing the process improves productivity but sacrifices security. In some cases, the request process is so cumbersome that it's rarely followed.

Ideally, all changes are recorded in logs, creating an accurate audit trail. The audit trail can be used during an audit or review to determine whether the approval process is being followed. When someone's account is granted administrative privileges, the audit trail provides information about who requested the change, who approved it, and who implemented it. It can also identify the source of unauthorized changes.

Administrator and other accounts with elevated privileges deserve the most attention when considering user entitlement. This includes controlling the number of users granted privileged access and limiting the number of users who can grant elevated privileges to others.

It's common to require administrators to use two accounts. Administrators log on with a regular account to perform typical day-to-day work; this account has limited privileges. They log on with the administrator account only when they need to perform administrative tasks.

True or false? All accounts deserve the same level of attention when managing user entitlement processes.

Answer: *False*. Administrative and other privileged accounts deserve more attention than regular user accounts. Accounts with privileges can cause the most damage to a company if misused.

EXAM TIP User entitlement involves more than just assigning privileges when an account is created. It also includes all the changes that occur during the lifetime of an account.

Access review and audit

Performing routine access reviews and audits helps an organization know whether security policies related to user accounts are being followed. This includes checks related to entitlement, provisioning, usage, and revocation.

One goal is to determine whether least privilege policies are being followed. A simple method is to periodically check the membership of groups that have a high level of privileges. For example, membership in administrative groups should be limited, and a routine audit will detect whether unauthorized individuals have been added.

Another method is reviewing logs that record user access and user provisioning. An organization will often define procedures for granting additional privileges to any user. A review of the logs used to track this process will determine whether the process is being followed or bypassed.

A security policy will typically specify whether accounts should be disabled or deleted for ex-employees, and a review can determine whether the policy is being followed. Cross-checking active accounts with an employee list can identify potential issues.

These checks can also discover unauthorized accounts. Imagine an administrator who is fired for cause but retains administrative access immediately after the exit interview. It takes less than a minute to create an account and give it full administrative privileges, including the ability to access the network from a remote location. Even if the ex-employee's account is disabled, 15 minutes later the damage is done.

A review can also determine whether administrators are using their accounts as dictated by the security policy. For example, administrators are commonly required to use two accounts—one for regular day-to-day work and the other for administrative purposes. Administrators might be tempted to use the administrative account all the time and never use the regular account. A review of the logs can identify whether administrators are using the regular accounts and how often they're using them.

True or false? When performing an access review, access to all data should be examined.

Answer: *False*. Only access to sensitive data should be examined. A review that examines access to all data will be extremely large and include data available to all users.

EXAM TIP Access reviews and audits should be repeated periodically, such as every six months or annually. Additionally, administrative or IT personnel should not do the review because the review is designed to inspect the processes and procedures of administrative and IT personnel.

MORE INFO Symantec published an interesting white paper, "User and Group Entitlement Reporting," which you can access here: http://eval.symantec.com/mktginfo/enterprise/white_papers/ent-whitepaper_entitlement_reporting_01-2006.en-us.pdf. It provides a good overview of user and group entitlement and includes some insight into what to look for when evaluating entitlement processes.

Can you answer these questions?

You can find the answers to these questions at the end of the chapter.

1. A user requires elevated privileges to perform a task once a week. What is the best way to assign these privileges?
2. What can be reviewed to determine whether an organization is complying with existing access control policies?

Objective 1.4: Identity and access provisioning lifecycle (e.g., provisioning, review, revocation)

The identity and access provisioning life cycle directly addresses the management of accounts from creation to deletion. When an account is first created, it is provisioned with appropriate privileges. During the useful lifetime of an account, these privileges are often modified and the account needs to be periodically reviewed to ensure that it has not been granted excessive privileges. When the account is no longer being used, such as when an employee leaves the company, it should be disabled as soon as possible and deleted when it has been determined that it is not needed.

Exam need to know...

- Understand issues related to provisioning of an account
  For example: What is permission creep?
- Understand review
  For example: Which accounts are the most important to review during the identity and access provisioning life cycle?
- Understand the importance of revocation
  For example: What should be done to a user account when an employee leaves the company?

EXAM TIP If you compare the information in the current Candidate Information Bulletin (CIB) with the previous one, you'll see that the "and access provisioning life cycle" is a new topic. Some of the concepts are similar, but it's worthwhile to recognize the importance (ISC)[2] is implying by adding the objective.

Provisioning

Provisioning refers to creating accounts and granting them access to resources. Role-based access control (or group-based) is often used to simplify management. Accounts are placed into groups that have defined privileges. As a best practice, all privileges are assigned via the role or group, and individual accounts are not granted privileges directly.

Some organizations use software to automate the provisioning process. For example, when an employee is hired, someone from the human resources department might enter the employee's information into an internal website application. This application is tied to a database and can automatically create the account and add it to the appropriate groups based on where the new employee will work.

Provisioning also occurs during the lifetime of an account when additional privileges are needed. For example, a salesperson assigned to the sales department needs privileges assigned to salespeople. If this person transfers to the IT department, the account is modified, adding privileges needed in the IT department.

Permission creep is a common problem that occurs when previously needed privileges are never removed. For example, someone who transferred from the sales department to the IT department no longer needs privileges assigned to salespeople. Without a procedure in place to remove unneeded privileges, many users progressively collect more and more privileges.

The use of roles or groups helps prevent permission creep. Users can be added and removed from the roles based on their current jobs, and they will automatically have the correct privileges.

Password policies and account lockout policies are often considered to be part of provisioning. Passwords are the weakest form of authentication, but strong password policies help ensure that users create strong passwords and regularly change them. They commonly include the following elements:

- **Password length** As tools to crack passwords become better and processor strength increases, the recommended length has also increased. An older recommendation is a password length of eight characters, but many security professionals now suggest a password length of 12 or more characters. Privileged accounts should be 15 or more characters.
- **Complexity** Passwords should have at least three of the four character types (uppercase, lowercase, numbers, and symbols). For the greatest complexity, passwords should include all four character types.
- **History** Users should be prevented from reusing the same password. A password history will often remember the last 12 or 24 passwords used by an account.
- **Maximum age** Users should be required to regularly change their passwords. Privileged accounts might be required to change their passwords every 30 days, and regular users might be required to change their passwords every 45, 60, or 90 days.

- **Minimum age**   This setting requires users to wait before they can reset their password again, and it is often set to one day. It prevents users from repeatedly resetting their password to bypass the history requirement and reuse the same password.

Account lockout policies lock out accounts when incorrect passwords are entered too many times. For example, they can be set to lock out an account after the user enters the wrong password five times in a 30-minute period. The account can be set to remain locked until an administrator unlocks it or for a set time such as 15 minutes. Some policies implement a delay after two or more failed login attacks and are very effective at preventing brute force attacks.

Password reset systems reduce costs by allowing users to reset their passwords without administrative intervention. Many require users to answer secret questions during a registration process, and these questions are later used to validate the user's identity before resetting the password. Attackers have used social engineering methods to learn these secrets and impersonate the user during the reset process. Password reset systems that communicate via email are less susceptible to these types of attacks.

True or false? Account de-provisioning is an important process that helps ensure that the principle of least privilege is enforced.

Answer: *True*. Account de-provisioning is the practice of removing privileges that are no longer needed and prevents permission creep.

> **EXAM TIP**   Processes should be in place to manage and audit the provisioning of an account throughout its lifetime. Without clear procedures, users often collect additional privileges and end up with more privileges they need. A review or audit can discover the problem.

> **MORE INFO**   Microsoft published a collection of technical papers titled Microsoft Identity and Access Management Series, which is available as a free download from here: *http://www.microsoft.com/download/details.aspx?id=17974*. This package includes a wealth of information that is valuable in understanding provisioning in centralized access control and federated identity management systems. The "Provisioning and Workflow" paper provides some excellent information on provisioning, the identity information life cycle, and different methods used to manage the identity life cycle.

Review

Accounts should be reviewed periodically to ensure that company policies are being followed. Privileged accounts are the most important to monitor so that misuse is quickly detected. It's often possible to detect suspicious activity by reviewing the logged activity of these accounts.

Groups are commonly used to grant privileges, and monitoring membership in these groups is also effective during a review. As a best practice, privileges should be granted only to a group or role rather than to an individual.

Monitoring group membership isn't the only review, though. The privileges assigned to the groups should also be periodically reviewed. Groups are assigned privileges based on job tasks. As additional job responsibilities are added, additional privileges can be added without removing unneeded privileges. Also, it's easy to focus only on permissions during a review, but the rights assigned to subjects should also be reviewed.

True or false? System logging is an effective measure used to identify misuse of privileged accounts.

Answer: *True*. System logs provide accountability as long as effective identification and authentication methods are used.

> **EXAM TIP** Periodic reviews are an important part of identity and access provisioning. They help identify whether security policies are being followed in the provisioning process and can be part of a formal access review and audit.

Revocation

Revocation of account access is a concept that most people understand, yet it is often not followed in practice. When an employee leaves the company, the account should be disabled as soon as possible. When the account is no longer needed, it should be deleted.

This is especially important for employees who have administrative privileges. There are more than a few stories where administrators were fired but retained access long enough to create unauthorized accounts with full administrative privileges that they later used to launch attacks.

Human resources (HR) departments can be valuable in keeping access control current. They know when employees are changing jobs and permissions should be changed, and they know when employees are being terminated and accounts should be revoked.

True or false? It is not necessary to immediately disable an account when an employee leaves after giving a notice.

Answer: *False*. It is just as important to disable accounts for employees who leave on good terms as it is to disable accounts for employees who have been fired.

> **EXAM TIP** Accounts should be disabled as soon as possible for employees who leave the company. When an employee is terminated for cause, a policy should be in place so that the account is disabled during the exit interview.

Can you answer these questions?

You can find the answers to these questions at the end of the chapter.

1. When is an account provisioned?
2. What can be used to review the provisioning process to determine whether the security policy is being followed?
3. When should an account be disabled?

Answers

This section contains the answers to the "Can you answer these questions?" sections in this chapter.

Objective 1.1: Control access by applying the following concepts/methodologies/techniques

1. A security policy. It has an overall goal of preventing loss of confidentiality, loss of integrity, and loss of availability of assets considered valuable by the organization.
2. A detective control. It is being used after an incident to discover what occurred. Audit logs can also be used for access reviews and audits, but, in that case, the logs are a preventive control.
3. An access control list (ACL) is directly associated with an object, and it lists subjects that can access it. An access control matrix combines a capability table with an ACL and lists all subjects and objects.
4. Biometric systems with a low crossover error rate (CER) are better than systems with a high CER. The CER identifies where Type 1 errors (false reject rates) are equal to Type 2 errors (false accept rates).
5. Service Assertion Markup Language (SAML) is used in federated identity management systems to share user information for single sign-on (SSO) between organizations in a federation.
6. A database view is a constrained user interface.
7. Audit trails require strong identification and authorization systems in place. If users are not uniquely identified or can easily be impersonated due to weak authorization, the data in the audit trails cannot be trusted.

Objective 1.2: Understand access control attacks

1. Input validation. Buffer overflow attacks are possible when users can enter unexpected data into a system and access normally inaccessible memory spaces. Input validation checks the data before it is used and prevents this type of attack.
2. All assets, including hardware, software, data, systems, and people, should be evaluated to determine their value.

3. The primary purpose of vulnerability scans is to evaluate the effectiveness of security controls in enforcing security policies.
4. Any person or organization can be a target of an APT. APTs are commonly thought to attack only military and government targets, but they have also targeted civilian organizations and individuals.

Objective 1.3: Assess effectiveness of access controls

1. Create a second account for the user and give it elevated privileges. Instruct the user to use this account only when it is necessary to complete the tasks requiring the elevated privileges.
2. A primary method of review includes the use of audit logs and audit trails. Another method is to identify who is assigned elevated privileges, such as by viewing membership in administrative groups.

Objective 1.4: Identity and access provisioning lifecycle (e.g., provisioning, review, revocation)

1. Accounts are provisioned when they are first created and throughout their lifetime. Provisioning occurs each time privileges are added, and de-provisioning is the process of removing privileges that are no longer needed.
2. Logs and audit trails are the primary method of reviewing the provisioning process. This requires the creation and proper management of logs and audit trails.
3. Accounts should be disabled as soon as it's known that they are not needed. When employees leave a company for cause, their account should be disabled during an exit interview.

CHAPTER 2

Telecommunications and network security

The Telecommunications and Network Security domain covers a wide range of topics directly related to networks, starting with core networking knowledge such as the OSI and TCP/IP models. Common networking hardware includes routers, switches, firewalls, and proxy servers, which can be used in local area networks (LANs), wide area networks (WANs), virtual LANs (VLANs), and virtual private networks (VPNs). Data is sent over both wired and wireless transmission media, and a variety of security measures are available to provide confidentiality, integrity, availability, and authentication for transmissions over both public and private networks. Networks are commonly attacked, and systems that are accessible on the Internet have the highest risk of attacks. Attackers can launch attacks against Internet-facing servers from anywhere in the world and use IP spoofing methods to hide their identity. It's essential to understand some basics about common network attacks along with common methods to protect against them.

This chapter covers the following objectives:

- Objective 2.1: Understand secure network architecture and design (e.g., IP & non-IP protocols, segmentation)
- Objective 2.2: Securing network components
- Objective 2.3: Establish secure communication channels (e.g., VPN, TLS/SSL, VLAN)
- Objective 2.4: Understand network attacks (e.g., DDoS, spoofing)

Objective 2.1: Understand secure network architecture and design (e.g., IP & non-IP protocols, segmentation)

For this exam objective, you need a basic understanding of the OSI and TCP/IP models, network topologies, and many of the commonly used protocols. The OSI and TCP/IP models provide a framework for networking, and protocols are mapped to specific layers in the models. IP is the primary protocol used in LANs, but you'll often run across non-IP protocols used in metropolitan area networks (MANs) and WANs.

Exam need to know...

- OSI and TCP/IP models
  For example: Which layer of the OSI model has two sublayers? Which layer of the OSI model provides logical addressing?
- IP networking
  For example: How does TCP initiate a session? What port does SNMP use?
- Implications of multilayer protocols
  For example: What types of circuits can be created with ATM? What is SONET?

OSI and TCP/IP models

The Open Systems Interconnection (OSI) model is a 7-layer model used to provide structure for networking protocols and devices. The Transmission Control Protocol /Internet Protocol (TCP/IP) model, also known as the Department of Defense (DoD) model, uses four layers.

Understanding the models is extremely useful for engineers designing protocols and hardware. They can focus on how to interact with the next layer without worrying about how all the layers work. Understanding the layers is also useful for technicians troubleshooting a network. If the problem can be isolated to a specific layer, it is easier to identify and repair the source of the problem.

True or false? The Address Resolution Protocol (ARP) operates on the Network layer of the OSI model.

Answer: *True*. ARP resolves Network layer addresses to link-layer addresses. Said another way, ARP resolves IP addresses to media access control (MAC) addresses.

Figure 2-1 shows the relationship between the OSI and TCP/IP models. Both models use encapsulation with higher levels (such as the Application layer) having smaller packet sizes. Each lower level adds additional information but encapsulates the data from higher levels.

> **MORE INFO** You might run across different names for the TCP/IP layers, but RFCs 1122 and 1123 are considered the definitive source by many people. RFC 1122 defines the communication protocol layers as the Link layer, Internet layer, and Transport layer. RFC 1123 defines the Application layer. You can read about both here: *http://tools.ietf.org/html/rfc1122* and *http://tools.ietf.org/html/rfc1123*. Wikipedia has an informative table showing the various ways the TCP/IP model has been described with different names and different numbers of layers: *http://en.wikipedia.org/wiki /TCP/IP_model*.

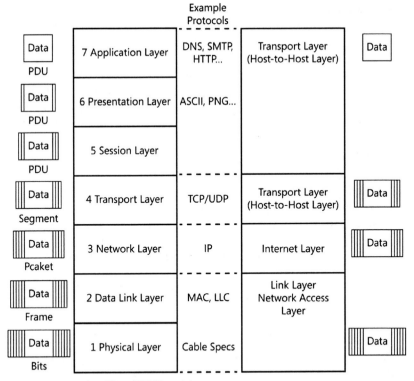

FIGURE 2-1 Comparing OSI and TCP/IP models.

- **Layer 7 (Application layer)** Network-based application protocols are defined here, such as Domain Name System (DNS), Simple Mail Transfer Protocol (SMTP), and Hypertext Transfer Protocol (HTTP). Data on this layer is called a protocol data unit (PDU).
- **Layer 6 (Presentation layer)** This layer includes protocols that define how data is represented using protocols such as American Standard Code for Information Interchange (ASCII) or Portable Network Graphics (png), and it also includes encryption protocols. Data on this layer is called a PDU.
- **Layer 5 (Session layer)** This layer manages the session between applications. Data on this layer is called a PDU.
- **Layer 4 (Transport layer)** Two important protocols on this layer are TCP, which provides a guaranteed connection, and User Data Protocol (UDP), which is connectionless using a best-effort delivery mechanism. TCP provides end-to-end error detection and correction. Data on this layer is commonly called a segment.

- **Layer 3 (Network layer)** IP is an important protocol on this layer, providing logical addressing with source and destination IP addresses. Data on this layer is commonly called a packet or a datagram.
- **Layer 2 (Data Link layer)** This layer uses physical addressing to provide reliable delivery of data. It includes the media access control (MAC) and logical link control (LLC) sublayers. Data on this layer is commonly called a frame.
- **Layer 1 (Physical layer)** The physical characteristics of media, such as cable specifications and how binary data is sent over the media, is defined on this layer. Data is sent as bits.

True or false? Both 802.2 and 802.3 are included in layer 2 of the OSI model.

Answer: *True*. LLC is a sublayer of layer 2 and is defined in 802.2. MAC is another sublayer of layer 2 and is defined in 802.3.

The TCP/IP model is less stringent about the specific lines between the layers, but in general, the Application layer maps to layers 5, 6, and 7 of the OSI model, the Transport layer maps to layer 4 of the OSI model, the Internet layer maps to layer 3 of the OSI model, and the Link layer maps to layers 1 and 2 of the OSI model.

> **EXAM TIP** You should have a basic understanding of the purpose of the different layers and be able to name or identify protocols used at specific layers. Figure 2-1 doesn't list all the protocols but instead just lists a few examples. Similarly, the layer descriptions are brief and do not explain the layers in any depth. If you want a more in-depth refresher, check out the Internetworking Technology Handbook on Cisco's DocWiki site: *http://docwiki.cisco.com/wiki/Internetworking_Technology_Handbook*.

Network topologies refer to the physical layout of networks on layer 1 of the OSI model. The primary topologies are as follows:

- **Star** Host devices on the network connect to a central device such as a hub or a switch. This is the most common network topology in use today.
- **Ring** Devices are configured in a circle or a ring. A token ring network passes a logical token to each device, and devices can send data only when they have the token. A token ring network often has a multistation access unit (MSAU) in the center to increase reliability. Fiber Distributed Data Interface (FDDI) rings use two rings.
- **Mesh** Each device is connected to every other device in the network, providing a high degree of redundancy for communications.
- **Bus** Devices are connected in a line, often with coaxial cable. Each end of the bus must be terminated, and a break in the bus connections stops all communication on the bus network. Whereas bus networks are rarely used in large networks, they are seeing a resurgence in home networks, connecting televisions, digital video recorders (DVRs), and gaming devices.

IP networking

Internet Protocol (IP) networking refers to the overall TCP/IP suite of protocols. IP is a specific protocol that operates on the network layer of the OSI model, and two versions are currently in use. IPv4 uses 32-bit addresses and is slowly being replaced by IPv6, which uses 128-bit addresses.

TCP and UDP are two core protocols within the TCP/IP suite that work at the Transport layer of the OSI model. TCP provides guaranteed delivery and UDP uses a best effort to deliver data.

True or false? UDP uses a three-way handshake when establishing a session.

Answer: *False*. UDP does not use a three-way handshake, but TCP does. A TCP three-way handshake provides assurances to both parties that a TCP session can be established. The originator sends out a SYN (synchronize) packet to start a TCP session. The receiving system responds with a SYN/ACK (synchronize/acknowledge) packet. The originator completes the handshake with an ACK packet.

After the TCP session is established, TCP provides error recovery by tracking the packets and ensuring that they are received. If a packet is not received, TCP can request a resend to receive it.

In contrast, UDP is connectionless and just sends the data to the receiving system. It doesn't establish a session or track the packets. UDP is commonly used in audio and video streams where some packet loss is acceptable and preferable to the overhead required to track the session.

Both TCP and UDP use ports to identify the protocol of the packet. The combination of the IP address and the port makes up a socket, or socket address. Well-known ports are numbered 0 to 1023 and assigned by the Internet Assigned Numbers Authority (IANA). Registered ports are numbered 1024 to 49151, and dynamic (also called private or ephemeral) ports are numbered 49152 to 65535.

True or false? SNMP is used to manage network devices, and it uses ports 49 and 69.

Answer: *False*. Simple Network Management Protocol (SNMP) is used to manage network devices, but it uses ports 161 and 162. Port 49 is used by TACACS, and port 69 is used by Trivial File Transfer Protocol (TFTP).

> **EXAM TIP** Traffic in and out of a network can be controlled by allowing or blocking specific port numbers. For example, to allow File Transfer Protocol, ports 20 and 21 are opened on a firewall. You should know the range of well-known ports and know many of the commonly used ports in the well-known range and the registered range. The following blog article covers common ports addressed for other exams, which also apply to the CISSP exam: *http://blogs.getcertifiedgetahead.com/ports-network-security-sscp-exams/*.

True or false? The IP address 172.17.5.2 is a valid public IP address.

Answer: *False*. This is a private IP address and should be used only in private networks. RFC 1918 identifies the private IP address ranges as 10.0.0.0 to 10.255.255.255, 172.16.0.0 to 172.31.255.255, and 192.168.0.0 to 192.168.255.255. Internet addresses are public IP addresses.

True or false? NAT hides clients in private networks.

Answer: *True*. Network address translation (NAT) translates private IP addresses to public IP addresses. Clients in private networks have private IP addresses, providing a measure of anonymity, and typically go through a proxy server, router, or firewall that has NAT installed.

Implications of multilayer protocols

Most network protocols follow the OSI or TCP/IP models and operate on a single layer. However, some protocols operate on multiple layers.

For example, Asynchronous Transfer Mode (ATM) is a non-IP protocol used to transfer voice, data, and video, and it operates on multiple layers. ATM uses three layers. The Physical layer corresponds to the Physical layer of the OSI model. The ATM layer and the ATM Adaptation layer correspond roughly to the Data Link layer of the OSI model.

ATM uses small 53-byte fixed cells rather than packets and sends the cells over a virtual circuit. A Switched Virtual Circuit (SVC) is created on demand, and a Permanent Virtual Circuit (PVC) is programmed in advance and can be thought of as a dedicated line for a customer.

True or false? Quality of Service (QoS) technologies built into ATM allow it to guarantee a specific bit rate for a PVC.

Answer: *True*. QoS technologies are built into ATM, and circuits can be created in one of four categories: constant bit rate (CBR), variable bit rate (VBR), available bit rate (ABR), and unspecified bit rate (UBR). Customers pay for the desired category based on their needs.

> *EXAM TIP* ATM technologies are used in Synchronous Optical Networking (SONET) and Synchronous Digital Hierarchy (SDH) fiber WANs. The speed and bandwidth of a SONET network over fiber cables is vastly superior to the speed and bandwidth over copper cables. For example, Optical Carrier 1 (OC-1) supports speeds up to 51.84 Mbit/s, and many faster versions are available. For example, OC-12 is 12 times the speed of OC-1 at 622.08 Mbit/s. In contrast, Frame Relay WANs are based on the T-carrier or E-carrier systems where a T-1 is 1.544 Mbit/s and an E-1 is 2.048 Mbit/s.

> *MORE INFO* Cisco has published a guide to ATM technology which provides excellent descriptions of ATM technologies, including a description of the ATM reference model. You can view it here: *http://www.cisco.com/univercd/cc/td/doc/product/atm/c8540/12_1/pereg_1/atm_tech/*.

Can you answer these questions?

You can find the answers to these questions at the end of the chapter.

1. What layer of the OSI model puts data onto the wire in a wired network?
2. Comparing TCP and UDP, which is more reliable and why?
3. What type of cable is used by a SONET WAN?

Objective 2.2: Securing network components

For this exam objective, you need to have a good understanding of the different hardware components used in a network. This includes the networking devices, such as routers and switches, as well as the transmission media. Wireless networks often present the highest level of risk for an organization, so it's important to know the basics of how to secure a wireless network.

Exam need to know...

- Hardware (for example, modems, switches, routers, wireless access points)
  For example: How are routers and switches commonly connected in a network? How do routers communicate known routes to each other?
- Transmission media (for example, wired, wireless, fiber)
  For example: Which media is immune to interference? What is an evil twin?
- Network access control devices (for example, firewalls, proxies)
  For example: What type of firewall will block or allow traffic by looking at more than a single packet? What is a bastion host?
- End-point security
  For example: What methods are used to protect smartphones? What are common methods used to protect desktop PCs?

Hardware (modems, switches, routers, wireless access points)

The primary hardware components used in a wired network are routers, switches, and patch panels. Computers in a network are connected to each other with a switch via a patch panel, and networks are connected together with routers via a patch panel.

Typically, cables run from computers to wall jacks, from wall jacks to a patch panel in a server room or wiring closet, and from a patch panel to a switch. Switches are connected to routers either directly in the same wiring closet or via patch panels. A wireless access point (WAP) provides a bridge to a wired network for wireless clients.

True or false? Physical security is an important consideration when securing routers and switches.

Answer: *True*. It's common to locate routers and switches in a server room or in a locked wiring closet. An attacker with direct physical access to these devices can connect a rogue wireless access point to capture the data. Switches support port mirroring, which mirrors all traffic through the switch to a specific port. Connecting the WAP to this mirrored port allows an attacker to capture the data and transmit it outside the organization's boundaries.

Routers have direct knowledge of networks to which they are connected but do not know of other networks by default. They use routing protocols to communicate known routes to each other. Routing Information Protocol 2 (RIP2) is a distance-vector protocol used in smaller networks by routers to share information. Open Shortest Path First (OSPF) is a link-state protocol used in larger networks to share information between routers and is able to build a topology database of the network.

Wireless routers are used in small networks to provide connectivity to the Internet. They include a wireless access point, switch capabilities to connect the computers in the network, routing capabilities to connect the internal network with the Internet, and a firewall. They support both wired and wireless clients.

Transmission media (wired, wireless, fiber)

The primary types of wired transmission in use are twisted-pair, fiber optic, and coaxial cables. Twisted-pair comes in both shielded twisted-pair (STP) and unshielded twisted-pair (UTP) versions.

Twisted-pair cable is the most common because it's relatively inexpensive and easy to work with. STP cable provides some protection against electromagnetic interference (EMI) and radio frequency interference (RFI).

> **EXAM TIP** A plenum space is an area between walls and ceilings through which cooled or heated air is forced. It's common to run cable through plenum spaces, but it's important to use only plenum-safe cable for fire safety. Plenum-safe cable is coated with a fire-retardant jacket and does not release toxic fumes when it's melted or burned.

Fiber optic cable has many benefits over twisted-pair cable, but it is the most expensive. Fiber cable runs can be longer than other cable types, have significantly greater bandwidth capabilities, and are not susceptible to interference.

True or false? Fiber optic cable is immune to EMI and RFI.

Answer: *True*. Data in fiber cable travels as light pulses, which are immune to EMI and RFI.

Wireless networks provide great flexibility, and many networks include both wired and wireless components. The speed and maximum distance of wireless networks are affected by many variables, such as antenna placement and interference. The primary wireless protocols are as follows:

- **802.11a** This has a maximum speed of 54 Mbps and uses a base frequency of 5 GHz.
- **802.11b** This has a maximum speed of 11 Mbps and uses a base frequency of 2.4 GHz.
- **802.11g** This has a maximum speed of 54 Mbps and uses a base frequency of 2.4 GHz.

- **802.11n** This has a maximum speed of 600 Mbps and uses base frequencies of 2.4 GHz and 5 GHz. 802.11n is the newest and uses multiple input multiple output (MIMO) antennas to achieve the higher speeds.

Security in a wireless network is especially important because it's easy for an attacker to use a directional antenna and capture transmissions. Wi-Fi Protected Access 2 (WPA2) provides the best security and supports Advanced Encryption Standard (AES) for encryption. Wired Equivalent Privacy (WEP) should not be used, and WPA is deprecated in favor of WPA2.

EXAM TIP WPA2 combined with authentication using an 802.1x authentication server provides the strongest security for wireless networks. Media access control (MAC) filtering provides only marginal security because it's very easy for an attacker to discover authorized MACs and modify the MAC on a computer. The default service set identifier (SSID) should be changed, but disabling SSID broadcast doesn't provide any real security. Disabling the SSID broadcast does hide the wireless network from some nontechnical users, but it doesn't hide it from a knowledgeable attacker.

MORE INFO Many people don't understand that disabling SSID broadcast doesn't provide any real security. The article "Myth vs. reality: Wireless SSIDs" provides a great explanation of why. You can read it here: *http://blogs.technet.com/b/steriley/archive/2007/10/16/myth-vs-reality-wireless-ssids.aspx.*

One of the risks in public wireless networks is an evil twin, which is a rogue access point configured with the same SSID as a legitimate WAP. If a user is tricked into connecting to an evil twin, the attacker hosting it can launch many different types of man-in-the-middle attacks, thereby collecting information from the victim.

True or false? Configuring a WAP in isolation mode isolates wireless users from each other.

Answer: *True*. Most wireless access points (WAPs) support isolation mode, which isolates clients from each other. Ideally, public WAPs should be configured to use isolation mode, but users typically will not know whether or not it's being used. As an additional protection measure, mobile devices connecting to public wireless networks should have a firewall enabled to block connection attempts from other users on the wireless network.

Network access control devices (firewalls, proxies)

A primary protection for a network is a firewall, which controls what traffic is allowed in or out of a network. Basic packet-filtering firewalls control traffic flow by filtering traffic based on IP addresses, ports, and protocols.

True or false? The last rule in a firewall will block all traffic not mentioned in a previous exception or rule.

Answer: *True*. This is also known as an implicit deny rule, which means all traffic is denied unless it was explicitly allowed by a previous rule.

EXAM TIP Exceptions or rules are created in an access control list (ACL), which allows traffic that meets the exception. For example, SMTP traffic can be allowed by creating an exception to allow port 25 traffic. Similarly, Internet Control Message Protocol (ICMP) traffic (used by ping and tracert) can be allowed by creating an exception for traffic using protocol ID 1.

MORE INFO IANA maintains a list of protocol numbers here: *http://www.iana.org /assignments/protocol-numbers/protocol-numbers.xml*. Some protocol IDs you might run across are ICMP -1, IGMP -2, IPv4 -4, TCP -6, and UDP -17. You can also view a full listing of all port numbers here: *http://www.iana.org/assignments/service-names -port-numbers/service-names-port-numbers.xml*.

Advanced firewalls control traffic by looking at more than just a single packet. For example, stateful inspection and application level firewalls can examine the entire session and allow or block traffic based on the context of the session.

Host-based firewalls are installed and enabled on individual hosts, providing an additional layer of protection. If the host has two network interface cards (NICs) or is a dual-homed computer, it's important to disable routing functions on the computer to prevent it from routing traffic on the network.

Networks that host web servers or other Internet-facing servers, such as email servers, commonly use a demilitarized zone (DMZ). A DMZ is usually configured with two separate firewalls, and ideally, to add defense diversity, each firewall is from a different vendor. For example, a DMZ with defense diversity might have a Microsoft-based firewall and a Cisco-based firewall. It is also possible to create a DMZ with a single three-pronged firewall using three network interface cards (NICs), although the configuration can be more complex.

MORE INFO Tony Northrup wrote an informative article on firewalls that is available on Microsoft TechNet. The article includes discussions of multiple ways of configuring a DMZ, including the use of a single three-pronged firewall and a DMZ with two separate firewalls hosting e-commerce servers. You can view it here: *http://technet .microsoft.com/library/cc700820.aspx*.

A bastion host is a hardened computer commonly placed within a DMZ. It will typically be a dedicated server with a single purpose. For example, it could be a reverse proxy server designed to distribute the load among multiple web servers in a web farm, or it could be a proxy server designed to provide central access to the Internet for internal clients.

True or false? Proxy servers can filter web content and block access to certain websites.

Answer: *True*. Proxy servers provide two important services. First, they can filter content and prevent users from visiting prohibited websites. Second, they reduce Internet bandwidth usage by caching website requests.

EXAM TIP System event managers (SEMs), security information management (SIM), and security information and event management (SIEM) are names for unified solutions designed to protect a network. They include an IDS or IPS component, collect data from multiple sources, secure the data in a central location, and can alert security personnel by using a variety of different methods.

True or false? Multiple network access control devices are used in a defense-in-depth strategy.

Answer: *True*. The most basic network access control device is a firewall. Additional devices include intrusion detection systems (IDSs), intrusion prevention systems (IPSs), proxy servers, antivirus software, and data loss prevention systems, which provide defense-in-depth when combined.

End-point security

End-point security refers to securing each device individually. This includes individual desktop or laptop computers and securing mobile devices such as tablets and smartphones. For example, a desktop PC would be kept up to date with current updates, have a host-based firewall enabled, and be running up to date antivirus software. It would also be locked down with basic security settings, such as changing defaults and disabling unneeded services and protocols.

True or false? Mobile devices with GPS enabled present a significant security risk if the device is lost.

Answer: *False*. Global Positioning System (GPS) can be used to locate a lost phone, so it is a security feature rather than a risk.

In addition to GPS, mobile devices can be protected by using other methods. Password-protecting the device with a screen lock password or PIN is an important first step. If the device holds sensitive data, encrypting helps prevent loss of confidentiality. If remote wipe is enabled, you can send a signal to the lost device that will delete all data. However, attackers also know about these methods. Attackers have been known to turn off stolen devices until they are in a shielded room similar to a Faraday cage. This blocks GPS and remote wipe signals. A simpler method thieves have used is to switch the device to "airplane mode" to stop it from receiving or transmitting any signals.

EXAM TIP Controls to restrict the use of USB devices on computers are often used to help prevent the spread of malware. An infected USB device can infect a computer when it is first plugged in, and this computer will then infect any other USB devices plugged into it. Additionally, some malware has a worm component and will attempt to spread through the network from the infected computer. Up to date antivirus software and host-based firewalls on each computer within the network help prevent the spread of USB-based malware.

Can you answer these questions?

You can find the answers to these questions at the end of the chapter.

1. When comparing link-state and distance-vector routing protocols, which one provides the most accurate view of a network?
2. What is the best way to secure a wireless network?
3. What is the primary way that traffic in or out of a network is controlled?
4. List four methods used to protect a tablet device.

Objective 2.3: Establish secure communication channels (e.g., VPN, TLS/SSL, VLAN)

Voice, audio, video, and digital data is routinely sent over networks, and this data is susceptible to sniffing attacks. Basic protocol analyzers (sniffers) can capture the data, and if it is sent in clear text, it can easily be viewed. An organization can implement security controls to reduce these risks for data sent both on internal networks and on the Internet. A core method of protecting sensitive data sent over the wire is by encrypting it.

Exam need to know...

- Voice (for example, POTS, PBX, VoIP)
  For example: Does VoIP transmissions need to be protected? What is RTP hijacking?
- Multimedia collaboration (for example, remote meeting technology, instant messaging)
  For example: Can teleconferencing meetings be captured by eavesdroppers? What is SPIM?
- Remote access (for example, screen scraper, virtual application/desktop, telecommuting)
  For example: What is a VM escape attack? How are VPNs protected when data is transmitted over a public network?
- Data communications
  For example: How can you ensure confidentiality for traffic between two systems? What benefits are provided by IPsec?

Voice (for example, POTS, PBX, VoIP)

Voice communications traditionally use the plain old telephone service (POTS). As long as someone has access to telephone lines, they can establish voice communications. POTS provides voice-grade or analog service, but many phone systems have been upgraded to digital. The public switched telephone network (PSTN) is primarily digital and includes telephone lines, microwave and satellite links, undersea cables, and cellular networks.

Many companies choose to implement a private branch exchange (PBX), which gives them much more control over phone lines used within the company. For example, a PBX can restrict users from making long distance calls. The PBX should be protected with physical security to reduce access, and standard digital security methods such as changing the defaults and regularly changing passwords should also be implemented.

Voice over Internet Protocol (VoIP) transmits audio over IP networks. Because the data is sent over IP, it can easily be captured with freely available tools such as Wireshark (*http://www.wireshark.org/*). Basic details of the call, such as the time, calling number, and called number, can be viewed easily. Wireshark also allows you to replay the audio.

VoIP media gateways provide a level of security for VoIP communications going over the Internet. They can limit the number of phone calls allowed through the gateway and block some attacks.

True or false? Attackers can insert media packets into video transmissions by using RTP hijacking.

Answer: *True*. Real-time Transport Protocol (RTP) is used for VoIP, and RTP hijacking injects media packets into a VoIP stream.

> **EXAM TIP** VoIP components should be protected with the same level of security as any IT device on the network. It's easy to think of VoIP phones as just phones, but because they are on the network, they are susceptible to network attacks. VoIP phones are actually less secure than regular phones because the data is transmitted over the Internet, which is more publicly accessible than telephone lines.

Multimedia collaboration (remote meeting technology, instant messaging)

Video conferencing (sometimes called teleconferencing) has become very popular with many organizations. It's possible to do face-to-face meetings without the time and cost related to plane trips and hotels.

True or false? Video conferencing using VoIP is immune to eavesdropping.

Answer: *False*. Any data sent over a network is susceptible to sniffing attacks, and this includes any type of VoIP traffic. Attackers who can tap into the session can capture the sound and video, effectively eavesdropping on the entire conference. In contrast, a conference held in a secure conference room with all attendees physically present is not subject to this risk.

Instant messaging (IM) tools such as Windows Live Messenger allow users to communicate with each other in real time over a network. The majority of the transmissions are single-line chats, but it's also possible to transmit files with IM tools. When public IM systems are used, users transmit messages and files over the Internet even when the users are in the same building. This makes IM transmissions susceptible to sniffing attacks. Additionally, attackers can impersonate others and send malware to unsuspecting users.

> **MORE INFO** The article "10 tips for using instant messaging for business" provides some basic security tips related to instant messaging: https://www.microsoft.com/business/en-us/resources/technology/communications/10-tips-for-using-instant-messaging-for-business.aspx.

Private IM systems provide administrators with more control over IM traffic. For example, Microsoft Lync Server (previously called Office Communications Server) includes all the features of instant messaging, supports conferencing and VoIP, blocks malware, prevents users from sharing inappropriate content, and can capture all conversations. This can be useful for end users in reviewing older conversations and is also useful in security audits and investigations.

> **EXAM TIP** SPIM is a form of spam over instant messaging channels. IM spam blockers are available to block SPIM when organizational users are connected to public IM systems. SPIM isn't a concern for private IM systems because internal users are less likely to spam each other.

Remote access (screen scraper, virtual application/desktop, telecommuting)

Remote access primarily refers to the ability of users to connect to an organization's network from a remote location. It can be done via dial-up or virtual private network (VPN) connections by users when they are traveling and when they telecommute. VPN connections are typically established over the Internet but can also be created using semi-private leased lines.

Dial-up connections commonly use Point-to-Point Protocol (PPP) over phone lines. PPP has very little built-in security and data is sent in clear text, making PPP connections susceptible to sniffing attacks, but the attacker needs access to the phone lines.

Tunneling protocols establish point-to-point connections and use additional measures to secure the connections. Some common protocols used in VPN connections are Internet Protocol security (IPsec), Layer 2 Tunneling Protocol over IPsec (L2TP/IPsec), Secure Sockets Layer (SSL), and Point to Point Tunneling Protocol (PPTP). VPN transmissions going over the public network should be encrypted. This includes host-to-gateway VPNs, where a single user connects to a VPN server, and gateway-to-gateway VPNs used to connect two separate offices.

> **EXAM TIP** VPNs commonly use multifactor authentication and encryption for increased security. For example, users are often required to use a smart card or token in addition to authenticating with a user name and password. Traffic within the VPN tunnel is encrypted to provide confidentiality.

True or false? PPTP is not used as often for VPNs as other tunneling protocols because of security weaknesses.

Answer: *True*. PPTP data is not encrypted in the initial negotiation, allowing the user name and hashed password to go over the Internet in clear text. Additionally, PPTP creates the encryption key from the user's password.

Remote access is also used by thin clients to run full desktops on a remote system or to run individual applications on a remote system. For example, Citrix has several products that allow users to run applications from their computers, which are hosted on a remote server.

True or false? A successful VM escape attack allows an attacker to gain control of all the virtual machines on a host computer

Answer: *True*. An attacker that has access to a virtual machine (VM) can launch a VM escape attack. If successful, the attacker gains full control over the host and can then control all the VMs. The primary protection is to keep the host machine and the VMs up to date with current operating system and application updates.

Screen scraping is the practice of harvesting data from a display screen. This can be done by reading the contents of display memory, executing the print screen function during a remote access session, or using code to record the data on the screen. Any remote access protocol that allows a remote user to view the desktop is susceptible to screen scraping without the user's knowledge.

> **NOTE** Some remote access protocols used for virtual desktops use a screen scraping technique to detect changes in the screen. They frequently scrape the screen and compare it to the previous scrape to detect changes. This can slow things down, and protocols more commonly use other techniques to detect changes.

Data communications

Data leakage is a concern to many organizations, but methods to prevent the loss of sensitive data are available. Any sensitive data sent over a public network or a private network should be encrypted to prevent a loss of confidentiality. IPsec, SSL, Transport Layer Security (TLS), and Secure Shell (SSH) are a few examples of encryption protocols used to protect data in motion.

IPsec uses both Authentication Header (AH) and Encapsulating Security Payload (ESP). AH is focused on authentication of the data origin and integrity, and ESP is focused on encryption. Overall IPsec provides confidentiality, data-origin authentication, connectionless integrity, and protection against replay attacks. It uses Internet Security Association and Key Management Protocol (ISAKMP) to establish security associations (SAs) between the endpoints.

The two modes used by IPsec are tunnel mode and transport mode. Tunnel mode encrypts the entire original IP packet and encapsulates it as part of a packet with a new IP header. In contrast, transport mode encrypts the data payload of the original IP packet but not its original header information.

> **MORE INFO** RFC 4301 (*http://www.ietf.org/rfc/rfc4301.txt*) formally defines IPsec. Windows-based systems have long supported IPsec, and the "IPsec: Frequently Asked Questions" page provides a good overview (*http://technet.microsoft.com /library/cc987611.aspx*).

The primary purpose of SSL and TLS is encryption, but they also provide authentication and data integrity. SSL and TLS depend on the use of certificates and a public key infrastructure (PKI) and use one of the Secure Hashing Algorithm (SHA) versions for integrity. SSL is still in use, but it is being replaced by TLS in many applications.

True or false? Encryption of data on the wire provides confidentiality between two endpoints.

Answer: *True*. Encryption is a primary control used to prevent loss of confidentiality.

> **EXAM TIP** Encryption helps prevent loss of confidentiality. This is equally true for data in motion and data at rest. There are many different protocols used to encrypt the data, but at the very core, encryption provides confidentiality. Similarly, hashing is commonly used to prevent loss of integrity. SHA-1 and SHA-256 are two common hashing protocols used with encryption protocols to provide integrity for data in motion and can also be used to provide integrity for data at rest.

True or false? Network-based data loss prevention can detect when a user is sending out sensitive data.

Answer: *True*. Data-loss prevention software is available and can perform content filtering on all data sent out by users. This helps prevent the loss of intellectual property, financial data, and sharing of personally identifiable information (PII).

> **MORE INFO** Ironport Systems (part of Cisco) published the "Data Loss Prevention Best Practices" document that defines the data-loss problem in more depth and provides several best practices to prevent data leakage. You can access it here: *http://www.ironport.com/pdf/ironport_dlp_booklet.pdf*.

Virtual local area networks (VLANs) are used to logically segment traffic on a network. For example, if an organization has two different departments named Sales and Marketing, they might want to segment the traffic so that broadcast traffic from either department is not sent to the other department. Switches are used to create VLANs, and each VLAN must include at least two ports.

True or false? Bluejacking is an attack against Bluetooth devices.

Answer: *True*. Bluejacking sends unsolicited messages over Bluetooth-enabled devices, and it can usually be prevented by ensuring that devices are in non-discoverable mode. Bluetooth is a wireless protocol used in small personal area networks (PANs). Devices are paired while they are in discoverable mode but should be switched to non-discoverable mode after being paired.

Can you answer these questions?

You can find the answers to these questions at the end of the chapter.

1. How should a PBX be protected?
2. What is SPIM and how can SPIM risks be reduced?
3. Name three protocols commonly used with VPNs.
4. What security benefits are provided by IPsec when it is used in a VPN?

Objective 2.4: Understand network attacks (e.g., DDoS, spoofing)

Years ago, it was news when a computer or network was attacked, but today security and IT professionals know that attacks are common and should be anticipated. It's important to understand the basic types of attacks along with some details of well-known attack methods. This section covers some DoS, DDoS, and spoofing attacks.

Exam need to know...

- DoS and DDoS
  For example: How many attackers are involved in a DoS attack? What are primary methods of protection against DoS and DDoS attacks?
- Spoofing
  For example: What type of attack is a hyperlink spoofing attack? What is modified in a DNS poisoning attack?

DoS and DDoS

A denial of service (DoS) attack is launched from a single attacker and tries to disrupt the target's ability to provide a service. For example, a successful buffer overflow attack against a web server might cause it to shut down or restart.

A distributed denial of service (DDoS) attack is a DoS attack launched from multiple attackers. For example, imagine that 1,000 computers simultaneously send repeated ping requests to a web server. If the web server tries to answer all the pings, it is unable to answer legitimate requests for webpages.

EXAM TIP You should be familiar with common attacks and know the difference between a DoS and a DDoS attack. A DoS attack is launched by a single attacker, and a DDoS attack is launched by multiple attackers. The attacking computers are usually not owned by the attacker but instead are remotely controlled.

True or false? A SYN flood attack attempts to flood a server with ICMP SYN packets.

Answer: *False.* A SYN flood attack sends a TCP SYN packet to a server to start a TCP three-way handshake, but it never sends the third packet (ACK) to establish a session. It's similar to a practical joker extending his hand out to you to shake hands, you extending your hand to him, and then him pulling his hand away.

EXAM TIP Servers reserve resources while waiting for the third packet, and it's common for a server to wait three minutes or more for the originator to send it. During a SYN flood attack, attackers often send hundreds of SYN packets using spoofed IP addresses and random ports. At some point, the server stops accepting new connection requests from both the attacker and from legitimate sources. Defenses include increasing the number of connections a server supports, decreasing the wait time, and using SYN cookies.

> **MORE INFO** If you want to review the basics of a SYN flood attack and some of the available protection methods, check out the article hosted by Cisco titled "Defenses Against TCP SYN Flooding Attacks" (*http://www.cisco.com/web/about/ac123/ac147/archived_issues/ipj_9-4/syn_flooding_attacks.html*).

True or false? The three elements of a smurf attack are an attacker, a victim, and an amplifying network.

Answer: *True*. A smurf attack is launched by an attacker as a broadcast ping with a spoofed source IP address. The spoofed source IP address is the victim, and the network that receives the broadcast ping is the amplifying network. The broadcast ping uses ICMP echo requests, and all clients on the amplifying network respond by sending ICMP echo replies to the victim.

A fraggle attack is similar to a smurf attack, but instead of using ICMP it uses User Datagram Protocol (UDP) ports 7 and 19. Chargen (character generator) is an older protocol (*http://tools.ietf.org/html/rfc864*) used to verify connectivity with a remote client. One system sends an echo request to the target using UDP port 19, and the receiving system responds, usually on UDP port 7, with a stream of characters. When the first system receives the characters, it knows the other system is operational.

In a fraggle attack, the attacker broadcasts a chargen request with a spoofed source IP address (the IP address of the victim). All systems on the network then flood the victim with a stream of characters.

True or false? Bot herders often launch DDoS attacks through command and control servers.

Answer: *True*. A bot herder controls botnets through command and control servers on the Internet. Malware infects computers, joining them to the botnet as zombies, and they periodically check in with the command and control server for directions. The bot herder can direct these zombies to launch attacks.

Attackers commonly have a suite of tools that they use to launch attacks. For example, Trinoo and Tribe Flood Network (TFN) are two tools that have been used to launch various DDoS attacks.

> **MORE INFO** Attack tools are constantly updated, just as any other software is regularly updated. Trinoo and Tribe Flood Network are older tools that were later combined into Stacheldraht. This Carnegie Mellon CERT note discusses some of the elements of Trinoo and Tribe Flood Network (*http://www.cert.org/incident_notes/IN-99-07.html*).

Core protection methods of protection against DoS and DDoS attacks include the following:

- **Disable unnecessary services and protocols** For example, if the chargen service is disabled, systems are not susceptible to a fraggle attack.
- **Close unneeded ports** If a port is closed, attacks using the related protocol are blocked. For example, systems with port 69 closed are immune to attacks using Trivial File Transport Protocol (TFTP).

- **Filter traffic at firewalls** Filters at boundary firewalls block unauthorized traffic.
- **Use IDSs and IPSs** These can detect attacks based on known attack signatures or behaviors and prevent the attack from succeeding. One way they can block an attack is by identifying attacking IP addresses and blocking traffic from these addresses.

Some organizations deploy honeypots to divert attackers. A honeypot looks like an attractive target to an attacker but does not have any data of real value. Some active IDSs modify routing rules to redirect attackers into isolated networks with only honeypots. At its core, the IDS is a detective control, but an active IDS that can respond to the attack by redirecting the attacker is a corrective control.

Spoofing

Spoofing is the practice of impersonating another system or person. In the context of networking, it most commonly refers to modifying the source IP address so that the IP packet looks like it's coming from another source. Spoofing also refers to modifying the MAC address or modifying an email From address.

True or false? A hyperlink spoofing attack is a form of a man-in-the-middle attack that can read traffic from secure Hypertext Transport Protocol Secure (HTTPS) sessions.

Answer: *True*. Normally, a user will establish a secure connection directly with a legitimate web server. In a hyperlink spoofing (also called web spoofing) attack, the user is tricked into establishing a secure session with the attacker's server, and the attacker's server establishes a secure session with the legitimate web server. Data is encrypted between the victim and the attacker and between the attacker and the web server, but it is not encrypted on the attacker's server, allowing the attacker to read the traffic from the session.

> *MORE INFO* The article titled "Hyperlink and Web Spoofing: Identifying and Defending Against Hacker Attacks" by Donna M. Shuart provides a good overview of IP spoofing with some in-depth explanations of hyperlink and web spoofing. You can download it here: *http://www.giac.org/paper/gsec/1483/hyperlink-web-spoofing-identifying-defending-hacker-attacks/102766*.

True or false? The attacker is easily identified in DoS and DDoS attacks.

Answer: *False*. Attackers commonly spoof the source IP address, so it's rarely easy to identify the original attacker. Additionally, many bot herders direct zombies in a botnet to launch attacks so that the zombies are innocent victims themselves. Many attacks can be stopped by blocking the source IP address, but that address is not necessarily the system launching the attack.

DNS spoofing poisons the cache of a DNS server, substituting malicious IP addresses for valid IP addresses in host address mappings. For example, a successful DNS poisoning attack can change the IP address for MSN.com to the IP address of a malicious website. When a user attempts to go to MSN.com, the user's system queries DNS for the IP address of MSN.com but instead receives the IP address for the malicious website.

In some cases, the redirected site is ready with a driveby download that infects the system as soon as the user visits. In other cases, the redirected site mimics the original site and the user is tricked into providing user credentials. The attacker can also infect a legitimate website with a cross-site script (XSS) and redirect the user to this site. When the user visits, the script runs within the user's browser.

EXAM TIP XSS attacks allow the attacker to execute malicious code on the user's system. This code allows the attacker to steal cookies and hijack web sessions.

Can you answer these questions?

You can find the answers to these questions at the end of the chapter.

1. Can an IDS detect a ping-of-death DoS attack?
2. Why would an attacker modify the source IP address?

Answers

This section contains the answers to the "Can You Answer These Questions?" sections in this chapter.

Objective 2.1: Understand secure network architecture and design (e.g., IP & non-IP protocols, segmentation)

1. The Physical layer sends the data out on the wire as bits. It receives data from the Data Link layer.
2. TCP is more reliable. It establishes a session with a three-way handshake and provides error recovery. UDP simply sends the data without checking to see whether the other system is operational, and it doesn't verify that all packets have been received.
3. Synchronous Optical Networking (SONET) wide area networks (WANs) use fiber optic cable.

Objective 2.2: Securing network components

1. Distance-vector routing protocols (such as OSPF) provide a more accurate view of a network than a link-state (such as RIPv2) routing protocol does. Link-state protocols only look at the number of hops. Distance-vector protocols evaluate many more variables.
2. Combining WPA2 with an 802.1x-based authentication server provides the strongest protection. It requires clients to authenticate before being granted access.
3. With firewalls. A basic packet filtering firewall controls traffic by examining individual packets. Advanced stateful inspection and application firewalls examine all the packets in a session.

4. Mobile devices such as tablets and smartphones can be protected by enabling GPS, enabling remote wipe, using a password, and encrypting data on the device.

Objective 2.3: Establish secure communication channels (e.g., VPN, TLS/SSL, VLAN)

1. With physical security to control access and with standard digital security methods used with any server.
2. SPIM is spam over IM channels. IM spam blockers and in-house IM systems can block SPIM.
3. L2TP/IPsec, IPsec, and SSL are often used with VPNs.
4. IPsec provides authentication of the data origin, integrity of the data, confidentiality through encryption, and protection against replay attacks. This is true when IPsec is used in a VPN and also when it's used within an internal network.

Objective 2.4: Understand network attacks (e.g., DDoS, spoofing)

1. Yes. A ping-of-death is a well-known attack, and an IDS can detect known attacks by using a database of known attack signatures. For most IDSs, you can substitute any known attack in the question and the answer is the same: yes, an IDS can detect known attacks.
2. Attackers modify the source IP address (called IP spoofing) to hide their identity.

CHAPTER 3

Information security governance & risk management

The Information Security Governance & Risk Management domain is a key domain within the CISSP candidate information bulletin and includes a longer objectives list than most other domains. You'll be expected to understand many elements of an organization's security program, with a focus on protecting information technology (IT) assets. This starts with a clear understanding of an organization's goals, mission, and objectives and then the development of security policies, standards, and procedures to support the mission of the organization. Risk management is an ongoing process that identifies asset values and then attempts to identify and prioritize risks to these assets. You'll find that the security triad of confidentiality, integrity, and availability is referenced in most CISSP domain materials, but this domain is one of the places it is addressed directly. Any risk management program will address the security triad principles either directly or indirectly.

This chapter covers the following objectives:

- Objective 3.1: Understand and align security function to goals, mission, and objectives of the organization
- Objective 3.2: Understand and apply security governance
- Objective 3.3 Understand and apply concepts of confidentiality, integrity and availability
- Objective 3.4: Develop and implement security policy
- Objective 3.5 Manage the information lifecycle (e.g., classification, categorization, and ownership)
- Objective 3.6: Manage third-party governance (e.g., on-site assessment, document exchange and review, process/policy review)
- Objective 3.7: Understand and apply risk management concepts
- Objective 3.8: Manage personnel security
- Objective 3.9: Develop and manage security education, training, and awareness
- Objective 3.10: Manage the Security Function

Objective 3.1: Understand and align security function to goals, mission, and objectives of the organization

A primary consideration related to information security governance and risk management is that it must be aligned with the organization's goals, mission, and objectives. Security controls aren't implemented just for security's sake. Instead, they are implemented to protect the organization's assets with a bias towards the vision of executive management.

Exam need to know...

- Align security function
  For example: What is the primary consideration when evaluating threats and vulnerabilities against an organization?

Align security function

Risk is the likelihood that a loss will occur. More specifically, it is the possibility that a threat will discover and exploit a vulnerability, causing a loss.

True or false? The main focus of an organization's information security program is to reduce risk.

Answer: *True.* An information security program organizes security activities with a main focus of reducing risk to its assets.

Different organizations view risks differently. Assets owned and valued by one organization might not be valued by another organization. Similarly, the vision and mission of organizations are rarely the same, and risk management programs are tailored to meet the needs of the organization.

With this in mind, an organization will often choose not to implement the same security control used elsewhere. This isn't necessarily good or bad. Instead, management weighs the overall cost of the controls against the risk based on the vision of the organization. They then make a decision to implement the controls or accept the risk.

True or false? A primary goal of any information security program is to protect the confidentiality, integrity, and availability of an organization's assets.

Answer: *True.* Preventing the loss of confidentiality, integrity, and availability are the basic tenets of any information security program and need to be reflected in an organization's security policy.

The methods used to protect confidentiality, integrity, and availability of assets vary and include people, processes, and technology. Technology cannot provide security by itself. The people within the organization implement security through effective processes and supporting technology.

EXAM TIP A security program's primary goal is to reduce risk to an organization's assets. A security policy (or security policies) is approved by management and should include their vision related to the goals, mission, and objectives of the organization. Members of the organization then use the security policy to develop standards, procedures, and guidelines.

MORE INFO National Institute of Standards and Technology (NIST) Special Publication (SP) 800-30, "Risk Management Guide for Information Technology Systems," provides excellent coverage of many risk management topics covered in this domain. It is only 55 pages and is highly recommended reading prior to taking the CISSP exam. You can download it from this page: *http://csrc.nist.gov/publications/PubsSPs.html*. Specifics related to confidentiality, integrity, and availability are included in Objective 3.3 in this chapter. Threats and vulnerabilities are explored in more depth in Objective 3.7 in this chapter.

Can you answer these questions?

You can find the answers to these questions at the end of this chapter.

1. What is the primary goal of an IT security program?
2. What is management's role in the development of a security policy?

Objective 3.2: Understand and apply security governance

Security governance refers to the different elements used to control and manage security within an organization. It includes various processes, personnel, and tools that an organization has in place to provide overall security. Organizations have a responsibility to ensure they are in compliance with all applicable laws and regulations and they will often use formal control frameworks to ensure that all elements are covered. Similarly, organizations have a responsibility to exercise due diligence to discover risks and due care to protect against known risks. Failure to exercise due diligence and due care can result in personnel within the organization being found negligent.

Exam need to know...

- Organizational processes (e.g., acquisitions, divestitures, governance committees)
  For example: What is the purpose of a security steering committee?
- Security roles and responsibilities
  For example: What is the role of senior management in regards to security? What are the responsibilities of data owners?
- Legislative and regulatory compliance
  For example: What is the relation between a security policy and local laws and regulations?

- Privacy requirements compliance
  For example: What is personally identifiable information (PII)? What is an organization's responsibility related to PII?
- Control frameworks
  For example: What is COBIT? What is COSO? What can an organization use to verify efforts to comply with the Sarbanes-Oxley Act?
- Due care
  For example: What is due care? How is negligence related to due care?
- Due diligence
  For example: What is due diligence? What is the relationship of due diligence with due care?

Organizational processes (e.g., acquisitions, divestitures, governance committees)

An effective security governance program ensures that security concerns are addressed in all organizational processes. It is much easier to ensure that a system or product is secure before it is implemented than to try to secure it afterwards. When developing security governance programs, the organization has a responsibility to balance risk and cost. The goal is to purchase and implement security controls that limit risk but also support the organization's overall mission.

True or false? A steering committee makes decisions on strategic and tactical issues for an organization.

Answer: *True.* A security steering committee makes security-based decisions to help define acceptable levels of risk for the organization.

A security-based steering committee includes input from multiple individuals throughout the organization. They meet regularly to review changes to the security policies and programs.

> **EXAM TIP** An organization might have multiple governance committees. For example, a steering committee includes several executives and provides overall direction for the organization, and an oversight or audit committee provides oversight for internal functions. These committees will typically report to the board of directors.

True or false? An audit committee must be external if it is used to validate an organization's internal controls.

Answer: *False.* An audit committee is an internal committee and it can gather data from either internal or external entities.

The goal of an audit committee (when used) is to evaluate the integrity of financial data, security controls, auditor performance, and compliance with applicable laws and regulations. They provide input to top level management, including the chief executive officer (CEO) and the board of directors.

Security roles and responsibilities

Security responsibilities are shared among multiple roles within an organization. Senior personnel typically have broad overall responsibility for assets. Roles that report to senior personnel are assigned more specific responsibilities based on the assets they are required to protect.

True or false? The primary role of senior management in conjunction with security is to ensure that a security manager is assigned.

Answer: *False*. Senior management has multiple roles well beyond assigning a security manager.

Some of the roles of senior management include the following:

- Ensuring that a security policy accurately reflects an organization's goals, mission, and objectives
- Providing public support for the security policy
- Identifying/approving definitions for data classifications
- Identifying data owners

True or false? Data owners are responsible for ensuring that data is properly labeled and for safeguarding their data.

Answer: *True*. Data owners are responsible for properly labeling, marking, and protecting their data. Labels and marks (such as secret or confidential) are based on the definitions used within the organization.

> **MORE INFO** Chapter 7, "Operations security," specifically addresses the marking, handling, storing, and destroying of sensitive information within the context of Objective 7.1. Senior management provides the definitions for the data, and data owners then ensure that the data is managed based on the sensitivity or value of the data.

There aren't any specific requirements for roles within an organization. However, some common roles include the following:

- Senior management has overall responsibility for all data under their purview.
- Data owners are typically management personnel, and they have overall responsibility for protecting their data, including ensuring that the data is properly labeled and marked.
- Data custodians are typically IT personnel, and they are responsible for maintaining and protecting the data.
- Users regularly access data during day-to-day tasks. They are responsible for following procedures for accessing and using the data.

> **EXAM TIP** Senior management has overall responsibility and accountability for all security. They can delegate their authority to lower roles to manage this responsibility, but they cannot delegate their responsibility.

Legislative and regulatory compliance

No matter where an organization is operating, it will fall under the jurisdiction of a government. Compliance with legislation and regulations is included within the security program throughout its lifecycle.

True or false? An organization's security strategies and practices must meet or exceed all governing regulations.

Answer: *True*. Organizations must ensure that their strategies and practices are in compliance with government regulations that apply to the organization.

Government regulations vary in different regions, so compliance can often be challenging when an organization operates simultaneously in multiple regions. However, ignorance of the laws or a lack of understanding of the complexities of the laws is not an acceptable reason for noncompliance. An organization always has a responsibility to understand and comply with all applicable regulations.

> **MORE INFO** Chapter 9, "Legal, regulations, investigations, and compliance," covers the Legal, Regulations, Investigations, and Compliance domain and addresses many specific legal and regulatory issues directly.

Privacy requirements compliance

Protecting data is important, especially when it comes to data governed by privacy laws. PII is formally defined in many regulations, and organizations have a responsibility to protect it.

True or false? PII includes any information that can be used to distinguish an individual, including name, date and place of birth, and mother's maiden name.

Answer: *True*. PII includes any information that can be used to identify an individual.

NIST SP 800-122 includes this definition of PII: "any information about an individual maintained by an agency, including (1) any information that can be used to distinguish or trace an individual's identity, such as name, Social Security number, date and place of birth, mother's maiden name, or biometric records; and (2) any other information that is linked or linkable to an individual, such as medical, educational, financial, and employment information."

> **EXAM TIP** PII is mentioned in privacy laws throughout the world. In the United States, the Privacy Act of 2005 addresses the protection of PII. Directive 2002/58/EC (the E-Privacy Directive) in the European Union also addresses data protection and privacy. These are just two examples, but there are many more laws around the world governing the protection of PII.

> **MORE INFO** NIST SP 800-122, "Guide to Protecting the Confidentiality of Personally Identifiable Information (PII)," provides in-depth definitions and examples of PII and discusses methods used to safeguard PII. NIST SP 800-144, "Guidelines on Security and Privacy in Public Cloud Computing," provides some different definitions for cloud computing and addresses many of the security and privacy issues related to protecting data used within cloud computing technologies. You can download both documents from this page: *http://csrc.nist.gov/publications/PubsSPs.html*.

True or false? Users should expect that any email they send by using an organization's resources will remain private.

Answer: *False*. A user who is sending email with an organization's resources (or any resources) should not expect it to stay private.

When any user sends an email by using any system, they have lost control of the email and shouldn't expect it to remain private. A recipient can send it to others or post its contents on a public website. Additionally, many organizations include privacy statements within their acceptable use policies that make it clear to users that their use of systems can be monitored.

True or false? The chief financial officer (CFO) is the person primarily responsible for protecting data within the organization.

Answer: *False*. The CFO is responsible for financial data but not for all data.

Some organizations have created a new "C" level position known as the chief privacy officer (CPO). This person is tasked with ensuring that all customer, employee, and organizational data is protected. In some cases, this person is an attorney and will report directly to the chief information officer (CIO) or the chief security officer (CSO).

Control frameworks

An IT control framework provides management with a set of tools, models, and best practices they can follow to implement and manage IT resources effectively. There are multiple frameworks available.

True or false? Control Objectives for Information and related Technology (COBIT) is a framework used to define goals for security controls.

Answer: *True*. COBIT is a set of good practices designed to provide a framework for IT management.

COBIT is a framework that can be used to define and manage any IT function, including IT security controls. COBIT version 5 is based on the following five principles:

- Meeting stakeholder needs
- Covering the enterprise end-to-end
- Applying a single, integrated framework
- Enabling a holistic approach
- Separating governance from management

NOTE The previous version of COBIT (version 4.1) had the following basic COBIT principles: business requirements drive investments in IT resources, IT resources are used by IT processes, IT processes deliver enterprise information, and enterprise information responds to business requirements. This is often drawn as a circle with business requirements at the top, IT resources on the right, IT processes on the bottom, and enterprise information on the left. Note that the process starts and ends with business requirements.

In addition to the five principles, COBIT version 5 lists the following seven enablers:

- Principles, policies, and frameworks are the vehicles to translate the desired behavior into practical guidance for day-to-day management.
- Processes describe an organized set of practices and activities to achieve certain objectives and produce a set of outputs in support of achieving overall IT-related goals.
- Organizational structures are the key decision-making entities in an enterprise.
- Culture, ethics, and behavior of individuals and of the enterprise are very often underestimated as a success factor in governance and management activities.
- Information is required for keeping the organization running and well governed, but at the operational level, information is very often the key product of the enterprise itself.
- Services, infrastructure, and applications include the infrastructure, technology, and applications that provide the enterprise with IT processing and services.
- People, skills, and competencies are required for successful completion of all activities and for making correct decisions and taking corrective actions.

EXAM TIP COBIT is a framework that is focused on implementing controls to support business requirements, and it includes principles, practices, tools, and models used for IT governance. The current version is COBIT 5 (released in June 2012) and is an upgrade from version 4.1.

MORE INFO ISACA (previously known as Information Systems Audit and Control Association but now known only by the ISACA acronym) manages and updates the COBIT standard. You can access a full set of resources on COBIT here: *https://www.isaca.org/cobit*.

True or false? The Committee of Sponsoring Organizations for the Treadway Commission (COSO) provides a framework to reduce fraud.

Answer: *True*. COSO provides guidance on enterprise risk management with the goal of improving organizational performance and governance and to reduce fraud.

EXAM TIP Many publicly traded companies that must comply with the Sarbanes-Oxley Act of 2002 (SOX) use COSO and/or COBIT as a framework.

MORE INFO You can read more about COSO, including its mission, vision, and history, here: *http://www.coso.org/aboutus.htm*. Additionally, it has several free white papers on enterprise risk management available here: *http://www.coso.org/guidance.htm*.

Due care

Due care refers to the steps an organization takes to protect itself against known risks. It is preceded by due diligence, which attempts to discover risks.

True or false? An organization that fails to meet at least the minimum required standards can be found to be negligent.

Answer: *True*. Negligence is defined as failing to use due care (or reasonable care) to meet at least the minimum standards.

An organization has a responsibility to exercise due care to meet normally expected requirements. Further, an organization has a responsibility to exercise due care to protect its assets even in the absence of formal standards.

> **EXAM TIP** Reasonable care is often defined in legal documents and interpreted in legal proceedings based on what a reasonable person would do. The reasonable person standard refers to an ordinary person and is used to determine whether conduct is rational and expected.

Due diligence

Due diligence refers to an organization's responsibility to expend a reasonable amount of time and effort to identify risks. This is followed by due care to protect the organization's assets against these risks.

True or false? An organization exercises due diligence to identify risks and due care to eliminate risks.

Answer: *False*. An organization exercises due diligence to identify risks, but it isn't expected to eliminate risks. Instead, it exercises a reasonable amount of effort to protect assets from known risks based on the goals, objectives, and mission of the organization.

Due care and due diligence are often discussed in the context of a fiduciary relationship where one entity trusts another. The trusted entity has a fiduciary responsibility to exercise due care and due diligence. For example, shareholders trust a board of directors to act in their best interests. Similarly, a board of directors trusts a CEO to act in the best interests of the board.

> **EXAM TIP** An organization has a responsibility to identify risks through due diligence. It then has a responsibility to address those risks using due care. The organization isn't expected to eliminate all risks and will often use different methods, such as risk assessments and cost benefit analysis, to decide how to manage various risks. After a serious security incident, management's judgment can be questioned with the claim that they are not exercising due care and due diligence. However, the existence of documentation such as risk assessments and actions taken to address various risks provides proof of both due diligence and due care related to IT security.

Can you answer these questions?
You can find the answers to these questions at the end of this chapter.
1. What is the purpose of a security-based steering committee?
2. An organization collects and stores user names, account names, passwords, and the maiden name of the user's mothers. Is this PII?
3. What are the basic principles of COBIT version 5?
4. What must an organization do to exercise due diligence?
5. What must an organization do to exercise due care?

Objective 3.3: Understand and apply concepts of confidentiality, integrity, and availability

Confidentiality, integrity, and availability are often referred to as the security triad. You might see it as the CIA triad or the AIC triad, but either way, the importance of protecting these three core security components cannot be overstated. Together they are often referred to as fundamental principles of security, and security programs have controls in place with the goal of preventing the loss of confidentiality, integrity, or availability.

Exam need to know...
- Confidentiality
  For example: What is the result if confidentiality is lost? What are two methods used to protect against the loss of confidentiality?
- Integrity
  For example: What is meant by loss of integrity? What is the most common method of checking integrity?
- Availability
  For example: What do hot and cold aisles have to do with availability? What is a single point of failure?

Confidentiality
Confidentiality refers to the ability of ensuring that unauthorized users cannot access data. That is, private or sensitive data remains private.

True or false? A primary method of ensuring confidentiality is by using encryption.

Answer: *True*. Encryption provides protection against the loss of confidentiality.

Confidentiality is also ensured through effective access controls. Users must first authenticate by using a user account that accurately identifies them. This account is then granted permissions to resources, such as files and folders, based on the user's need to know. Accounts without permissions are not granted access to resources, helping to prevent a loss of confidentiality.

EXAM TIP Encryption and access controls help prevent the loss of confidentiality. Encryption can be applied to data at rest (as files) and data in transit (traveling over a network). Access controls ensure that personnel are properly identified and authenticated, and permissions control their access based on their proven identity. Confidentiality is lost when unauthorized individuals can access data.

MORE INFO Chapter 1, "Access control," covers access controls, including identification and authentication topics. Chapter 5, "Cryptography," covers many different cryptography concepts, including encryption methods used to protect data at rest and data in transit.

Integrity

Integrity controls help prevent unauthorized modifications to data and ensure that unauthorized modifications are discovered. They also help prevent or detect corruption of data.

True or false? Hashing methods are frequently used to detect corrupted or unauthorized modified data.

Answer: *True*. Hashing methods are commonly used to check the state of data at two different times, such as before a message is sent and after it is received.

In the simplest terms, a hash is a number, and it's normally displayed as a hexadecimal number. The hash (the number) created by a hashing algorithm will always be the same as long as the original data is the same.

Hashing methods can be used to verify that messages or data files have not been modified. For example, if the hash of a file is 1A2B3C on Monday but it is 4D5E6F on Friday, the file is no longer the same. If the file were the same, the hash on Friday would be the same as the hash on Monday. If the modification was not authorized (such as an infection of a system file by a virus), the comparison verifies that the file has lost integrity.

Data can also lose integrity accidentally. For example, if a database programmer attempts to increase the price of all products by five percent but incorrectly uses .50 as the multiplier instead of .05, all the prices are increased by 50 percent instead of 5 percent. Even though this is an accident, the result is the same. The data has lost integrity.

EXAM TIP A primary goal of integrity is to protect against the unauthorized modification of data. Hashing techniques are used to detect modifications.

MORE INFO Chapter 5 covers various hashing methods and algorithms, such as Message Digest 5 (MD5), and different Secure Hashing Algorithm (SHA) versions.

Availability

Availability ensures that an organization's resources are available when needed. Some resources, such as e-commerce websites, are expected to be available 24 hours a day every day of the year. Other resources, such as an internal file server, are needed only during regular working hours. There are multiple security controls that help support availability.

True or false? Hot and cold aisles used in a large data center contribute to availability.

Answer: *True*. Hot and cold aisles are used to improve the performance of heating ventilation air conditioning (HVAC) systems in data centers.

If an HVAC system fails, servers can overheat and fail, so HVAC systems directly contribute to availability. When hot and cold aisles are not used, some server bays are typically hotter than others, potentially reducing their reliability.

Any systems or services that ensure that a system continues to operate contribute to a system's availability. This includes any system that contributes to fault tolerance or system resilience by removing single points of failure. For example, if power is lost, an uninterruptible power supply (UPS) provides short-term power and a generator provides long-term power. Redundant array of independent disks (RAID) contributes to availability if a disk drive fails, and failover clusters contribute to availability if a server fails.

> **EXAM TIP** Assets should be available to users when they need them, based on the mission of the organization. Some systems need to be up and operational 24 hours a day every day of the year, and the cost of additional redundant systems is justified. These same redundant systems are not necessarily justified for systems that are not used as often. Redundant systems attempt to remove single points of failure.

> **MORE INFO** The Operations Security domain (covered in Chapter 7) addresses system resilience and fault tolerance in Objective 7.7. The goal is to reduce or eliminate single points of failure, increasing overall availability.

Can you answer these questions?

You can find the answers to these questions at the end of this chapter.

1. What are the two primary methods of preventing loss of confidentiality?
2. A system file has been modified by a virus, but the file name and location is the same. What can be used to detect the modification?
3. What do a RAID, an UPS, and a failover cluster all have in common?
4. Backup tapes of sensitive data were stolen after they were created. These tapes were not encrypted. Considering confidentiality, availability, and integrity, what has the organization lost?

Objective 3.4: Develop and implement security policy

A security policy is a written document that provides overall strategic direction for an organization's security. From this, tactical methods are used to implement the strategy with standards, baselines, procedures, and guidelines. Another way of looking at this is that security policies are created to meet the vision of the organization. Security policies comply with laws, regulations, and external standards, and they drive the creation of internal standards when necessary or desired. Guidelines are used to help in the creation of detailed procedures.

Exam need to know...

- Security policies
  For example: What type of information is included in a security policy? Who provides input to security policies?
- Standards/baselines
  For example: What is an example of an external standard? What is a security baseline?
- Procedures
  For example: What is the difference between a procedure and a standard? Who are the primary users of procedures?
- Guidelines
  For example: What is the difference between a guideline and a procedure?
- Documentation
  For example: How often should a security policy be updated? Who should have access to the security policy documentation?

Security policies

Security policies are written documents that provide an organization overall direction relating to security. There aren't any restrictions or requirements for the length of a security policy, but it's not uncommon to see a security policy as long as 50 pages for a large enterprise. In some organizations, a primary security policy is created and it references additional security policies. This makes it easier to update any of the policies when necessary.

True or false? Security policies provide the details for personnel about how to implement and maintain various security procedures.

Answer: *False*. Security policies are high-level documents that identify the overall goals and vision of the organization.

Senior management ultimately decides what is in the organization's security policies. With this in mind, it's essential that security policies are in line with senior management's goals and vision for the company. The concepts contained in the security policies are then used to create standards, procedures, and guidelines.

True or false? Security policies are written with input from several entities within the organization, including human resources and legal personnel.

Answer: *True*. Security policies require input from multiple parts of the business, not just IT and security personnel.

Security policies define the overall goal of security, but they do not dictate specific goals that should be met. For example, a security policy might mandate strong authentication when users authenticate in-house and two-factor authentication when users connect via a virtual private network (VPN). The actual authentication methods used are not specified, but the authentication goals are described.

> **EXAM TIP** A security policy provides overall direction for security within an organization. It requires input from multiple entities within the organization, but senior management is responsible for the information in the policy and ensuring that the policy is followed. A security policy should always follow all laws and regulations.

> **MORE INFO** The SANS Institute maintains a list of information security policy templates that are available here: *http://www.sans.org/security-resources/policies/*.

Security policies are often divided into the following types:

- **Functional** These are sometimes called issue-specific security policies and outline details with specific security issues. For example, an organization might decide to restrict the use of USB flash drives and create a policy informing users about what is allowed and what isn't allowed.
- **System** These policies outline requirements for specific systems or networks. For example, a policy might dictate that all servers and end-user computers are protected with antivirus software. Similarly, a policy might dictate the use of network-based data loss protection systems to prevent data leakage.
- **Regulatory** These are driven by government or industry regulations. Any laws that apply to an organization are included in regulatory policies.
- **Advisory** These provide advice and requirements for internal users. For example, an acceptable usage policy is an advisory policy targeting users.
- **Informative** These policies provide information and education for employees. They are primarily used to help employees understand certain topics so that they can contribute to the organization's overall security.

Standards/Baselines

A standard is a group of mandatory requirements that must be met to comply with the standard. They might be identified as actions or rules that must be followed, but the key is that they are mandatory. Internal standards are designed to support compliance with a policy or policies. In contrast, security policies are often written with compliance to the external policy in mind.

True or false? Security standards are similar to guidelines.

Answer: *False*. Security standards identify mandatory requirements, but guidelines are recommendations for security best practices.

Standards might be specified internally or externally. For example, laws and regulations are examples of external standards that an organization must follow. Similarly, the Payment Card Industry Data Security Standard (PCI DSS) is an external standard that an organization must follow to process certain credit card transactions. An organization can also create its own standards and mandate their use throughout the organization.

> **MORE INFO** Chapter 6, "Security architecture & design," mentions the PCI DSS in Objective 6.2. PCI DSS includes 12 specific requirements that retailers agree to abide by before being approved to process credit card transactions.

True or false? A security baseline provides a minimum starting point for security.

Answer: *True*. A baseline is a starting point, and a security baseline is the starting point for security. Security can be strengthened from the baseline but should not be weakened.

Configuration management practices often include the use of images to deploy operating systems. The image is created with an operating system, applications, and a security baseline. After testing, the image is captured and can then be deployed to multiple other systems. Each system starts with the same configuration, including the same security baseline.

The security baseline is typically created after a significant amount of research and testing. Additionally, as things change, the security baseline is updated to reflect new capabilities and features and also to protect against newer threats.

For example, the United States Air Force worked with Microsoft and National Security Agency (NSA) personnel to create the Standard Desktop Configuration (SDC) as a standard image for all computers. This was later adopted by the Office of Management and Budget (OMB) as the Federal Desktop Core Configuration (FDCC) for all government computers.

> **EXAM TIP** A standard is a mandatory requirement that might be external or internal. A security baseline is a starting point that identifies the minimal security requirements for a system, and the baseline might be enhanced to increase the security.

> **MORE INFO** You can read more about the FDCC project here: *http://www.microsoft.com/industry/government/federal/fdccdeployment.aspx*. The page also includes links to white papers, webcasts, and other resources related to the development and ongoing maintenance of the FDCC image.

Procedures

Procedures are step-by-step actions performed for a specific outcome or goal. They are often used by IT personnel to assist in completing various tasks such as backup or restore operations.

True or false? Procedures are part of the implementation of a security policy.

Answer: *True*. Procedures identify how specific tasks can be completed, and these tasks are completed in response to direction from the security policy.

For example, a security policy might dictate that certain types of data must be backed up regularly and that these backups must be verified. IT personnel can then create step-by-step procedures to complete backups. To test the backups, IT personnel can create procedures to periodically restore the data and verify that the backup procedures are working.

> **EXAM TIP** Procedures provide the step-by-step instructions for various tasks. They ensure that personnel performing the tasks have clear direction about how to complete the tasks. Procedures also help an organization remain in compliance with any governing standards or regulations and with its own internal security policies.

Guidelines

Guidelines are suggestions or recommendations that are often used to identify best practices. Security policies will often reference a guideline or a standard as a best practice, but this is as a recommendation.

True or false? Guidelines are directive in nature.

Answer: *False*. A security policy is directive in nature, but guidelines are recommendations.

An organization that supports banking over the Internet might state in the security policy that guidelines from the Federal Financial Institutions Examination Council (FFIEC) should be consulted. Personnel within the organization have a responsibility to check these guidelines but not necessarily follow them completely.

> **EXAM TIP** Guidelines are often used to provide additional guidance to personnel within the organization in support of a security policy. For example, strong authentication might be mandated in the security policy. The security policy might refer to a reference as a guideline for strong authentication. A guidance can recommend the use of 15-character passwords for anyone with administrative permissions and 8-character passwords for regular users.

> **MORE INFO** The FFIEC has published a white paper titled "Authentication in an Internet Banking Environment," which you can access here: *http://www.ffiec.gov/pdf/authentication_guidance.pdf*.

Documentation

Security policies, standards, baselines, procedures, and guidelines need to be well documented within the organization. This documentation needs to be easily understood and accessible to all personnel who have responsibilities from the documentation.

True or false? Security policies should be created once, and only standards, baselines, procedures, and guidelines should be updated.

Answer: *False.* Security policies should be regularly reviewed and updated when necessary.

All documentation created internally in support of security policies should be regularly reviewed to ensure that it continues to meet the needs of the organization. Changes should be well documented by using appropriate change procedures for the organization, and all appropriate personnel should be kept informed of any changes.

> **EXAM TIP** Security policies can be kept in a single large security policy or as one primary policy with multiple supporting policies. However, it's never recommended to include the standards, baselines, procedures, and guidelines within the same document. These should be separate so that they can be easily modified by the appropriate personnel when needed. Senior management needs to approve changes in security policies, but they do not need to approve changes in most of the supporting documentation.

Can you answer these questions?

You can find the answers to these questions at the end of this chapter.

1. What external elements must be considered when developing a security policy?
2. What does a security baseline provide for an organization?
3. Administrative personnel have created a document showing how to perform backups and another document showing how to restore data when necessary. What type of document is this?
4. Who can approve changes to the security policy?

Objective 3.5: Manage the information lifecycle (e.g., classification, categorization, and ownership)

Information is often one of the most valuable assets an organization owns, and it's extremely important for an organization to protect it. Sensitive information must be protected against the loss of confidentiality, the loss of integrity, and the loss of availability. An effective information lifecycle management program includes clear definitions of data classifications. This helps personnel within the organization effectively balance resources to protect the most valuable data.

Exam need to know...

- Manage the information lifecycle
  For example: What's the purpose of classifying data? Who is responsible for defining data classifications?

Manage the information lifecycle

Every organization has secrets that it protects. For example, an organization will often consider proprietary data to be sensitive and prevent its release outside the company. It restricts access to the data to only specific individuals within the company and takes extra precautions to ensure that it is not released outside the company. In contrast, public data is not protected from disclosure but instead is publicly available. When data is properly classified, personnel recognize the value of the data that they are handling.

True or false? Administrators are responsible for defining information classifications.

Answer: *False*. Senior management (not administrators) is responsible for defining information classifications.

An organization can decide to use classification levels of confidential, private, sensitive, and public. Senior management is responsible for ensuring that each of these levels is clearly defined. For example, they might identify proprietary information as confidential and products in their sales catalog as public.

In addition to defining classifications, senior management is also responsible for identifying data owners. Data owners evaluate the data they own and mark or label it according to the definitions. Data owners are also responsible for taking precautions to protect the data. Data with higher classifications, such as confidential, warrants additional steps to protect it, especially against the loss of confidentiality. In contrast, public data is meant to be public. It doesn't need to be protected against loss of confidentiality, but personnel will take steps to protect it against loss of integrity or loss of availability.

> **EXAM TIP** Senior management is responsible for defining data classifications and assigning data owners. Data owners must take steps to protect their data from loss of confidentiality. IT personnel are responsible for protecting the data against the loss of availability.

True or false? Properly classifying data helps ensure that security professionals can protect all data equally.

Answer: *False*. Classifying data helps security professionals recognize the valuable data and ensure that appropriate resources are available to protect it. In contrast, public data doesn't require as much protection.

> **EXAM TIP** Common classifications of data used in public organizations from highest classification to lowest are as follows: confidential, private, sensitive, and public. Governments often use labels such as top secret, secret, confidential, and unclassified. When classification procedures are in place and followed, they help prevent loss of confidentiality.

Stringent security controls are justified when protecting data with higher classifications. When data is properly classified and labeled, it's clear to security professionals when extra precautions are needed.

MORE INFO Chapter 7 includes information about the proper marking, handling, storing, and destroying of sensitive information. Policies identify the definition of the different classifications, and then procedures are used to properly mark and handle this information. The overall goal is to prevent unauthorized disclosure of data, with a heavier emphasis on protecting data with higher classifications.

True or false? One of the goals of classification procedures is to protect the confidentiality of data.

Answer: *True*. When data is properly classified and labeled, users can easily identify highly valuable data and are less likely to mishandle it. This helps prevent the unauthorized disclosure of the data and contributes to overall confidentiality.

Can you answer these questions?
You can find the answers to these questions at the end of this chapter.
1. What is senior management's role related to data classification?
2. What is the role of a data owner?
3. What is the primary benefit of using data classifications?

Objective 3.6: Manage third-party governance (e.g., on-site assessment, document exchange and review, process/policy review)

In addition to internal governance, many organizations need to be concerned with third-party governance performed by outside entities. The requirements are based on the function of the organization and the rules and regulations that apply to it.

Exam need to know...
- Third-party governance
  For example: What is third-party governance? What can an organization provide to show it is in compliance with third-party governance?

Third-party governance
Governance is the act of governing, or what a government does, but the term takes on different connotations in the context of IT governance, corporate governance, and third-party governance. Corporate governance comprises the processes an organization uses to control and monitor compliance with its vision and goals. IT governance refers to the processes used to manage IT systems and ensure that the IT investments provide clear value for the organization. Third-party governance refers to the external processes used to assess an organization's compliance with established requirements.

True or false? A medical organization that provides health care in the United States must abide by the Health Insurance Portability and Accountability Act (HIPAA), but this does not apply to organizations that do not provide health care.

Answer: *False*. HIPAA applies to all organizations in the United States that handle any type of health information.

Health information is defined as any information that includes past, present, or future health, including physical health, mental health, or physical condition. It also includes any past, present, or future payments for health care. This information might be created, received, or maintained by the following entities:

- Insurers
- Employers
- Health plans
- Health care providers
- Public health organizations
- Schools, colleges, and universities

EXAM TIP Organizations that fall under HIPAA or any outside rule or regulation have a responsibility to ensure that they are in compliance with the regulation. This is done by regular assessments to verify compliance and identify risks, followed by the creation and implementation of compliance plans. These assessments can be done internally, by external personnel, or by a combination of both.

MORE INFO HIPAA is mentioned in Chapter 9, which covers the Legal, Regulations, Investigations, and Compliance domain.

True or false? An organization found to be noncompliant with a third-party rule or regulation can mitigate the penalties if it has documentation showing regular reviews.

Answer: *True*. The existence of documentation that shows an honest effort by an organization to comply with rules and regulations is often considered when penalties are assessed.

If an organization regularly completes compliance assessments and reviews and if documentation shows that they have taken steps to correct any deficiencies, the governing agency often takes this into consideration. In contrast, an organization that is found to be noncompliant and doesn't have any type of compliance program in place will likely face heavier penalties.

Penalties vary based on the governing agency but are often monetary penalties, more stringent review and assessment requirements, or loss of specific privileges. For example, PCI DSS can require an organization to pay between US$5,000 and US$100,000 per month for penalty violations. It can also increase transaction fees or terminate the business relationship, preventing the business from processing credit card transactions.

EXAM TIP It's extremely important for an organization to document all of its processes related to third-party governance. When requested, it can provide this documentation for review to validate its compliance efforts. Without documentation, it's as if the company is not making any effort to comply with the rules or regulations.

MORE INFO The IT Governance Institute (*http://www.itgi.org/*) was established as a resource to help organizations develop IT governance programs to better manage IT resources. Its site has links to multiple resources; ITGI, COBIT, and ISACA are all linked to each other.

True or false? The Sarbanes-Oxley Act is an example of a third-party governance requirement.

Answer: *True*. Sarbanes-Oxley applies to all companies required to file reports with the United States Securities and Exchange Commission.

Under Sarbanes-Oxley, company executives such as chief executive officers (CEOs), chief financial officers (CFOs), and auditors are required to personally verify the accuracy of financial data reports. The goal is to reduce the incidence and severity of fraud in publicly traded companies.

Can you answer these questions?

1. What is the difference between IT governance and third-party governance?
2. Who must abide by the United States HIPAA law?
3. Who must abide by the Sarbanes-Oxley Act?

Objective 3.7: Understand and apply risk management concepts

Risk management is an ongoing process. It starts by identifying valuable assets within an organization and then regularly identifies threats and vulnerabilities related to these assets. A risk assessment is a point-in-time review that attempts to identify the potential for any threat to exploit a vulnerability resulting in a loss. Management decides whether they want to mitigate, avoid, transfer, or accept a risk. Risk mitigation is performed by selecting and implementing countermeasures (also called safeguards or controls) that can reduce the risk. The overall goal is to protect an organization's valuable tangible and intangible assets.

Exam need to know...

- Identify threats and vulnerabilities
  For example: What is the difference between a threat and a vulnerability? What are some natural threats that an organization should consider?
- Risk assessment/analysis (qualitative, quantitative, hybrid)
  For example: What type of risk assessment uses a monetary value to prioritize risks? What type of risk assessment uses the opinion of experts to prioritize risks?

- Risk assignment/acceptance
  For example: What is the difference between risk transference and risk avoidance? What is residual risk, and who is responsible for losses related to residual risk?
- Countermeasure selection
  For example: How is annual loss expectancy calculated? If a countermeasure is effective, what quantitative risk measurement will it reduce?
- Tangible and intangible asset valuation
  For example: What is the difference between tangible and intangible assets? Are intangible assets included as part of an organization's net assets?

Identify threats and vulnerabilities

As mentioned previously, risk is the probability of a threat exploiting a vulnerability. With this in mind, it's important to have a basic understanding of threats and vulnerabilities.

True or false? A vulnerability is a weakness in any part of an IT system.

Answer: *True.* Vulnerabilities are weaknesses, and these weaknesses can be in any part of an IT system.

NIST SP 800-30 formally defines a vulnerability as "a flaw or weakness in system security procedures, design, implementation, or internal controls that could be exercised (accidentally triggered or intentionally exploited) and result in a security breach or a violation of the system's security policy." Here's a simpler definition that fits most situations: a vulnerability is a weakness that can result in a security incident.

Documentation is extremely useful when attempting to identify vulnerabilities. This includes previous risk assessment reports, audit and review reports, public vulnerability advisories, and internal security incident reports.

True or false? When evaluating threats within risk management, only human threats need to be considered.

Answer: *False.* Human threats, natural threats, environmental threats, and operational threats all need to be considered as a part of risk management.

Human threats include any potential threats caused by humans either deliberately or accidentally. Malicious individuals perform a wide variety of malicious attacks. Users can also perform unintentional actions that adversely affect the confidentiality, integrity, and availability of an organization's assets.

Natural threats include events such as floods, hurricanes, tornadoes, earthquakes, and lightning storms. Environmental threats include events that result in long-term power failure or a release of chemicals or other pollutants. Operational threats are associated with automated or manual processes within the organization.

NIST SP 800-30 provides the following definitions for a threat and a threat-source:

- A *threat* is "the potential for a threat-source to exercise (accidentally trigger or intentionally exploit) a specific vulnerability."

- A *threat-source* is "either (1) intent and method targeted at the intentional exploitation of a vulnerability or (2) a situation and method that may accidentally trigger a vulnerability."

EXAM TIP One of the goals during a risk assessment is to gather a list of threat-sources that have the potential to exploit known vulnerabilities. This can be used to create a list of vulnerability/threat pairs. A threat pair is a threat matched with one or more vulnerabilities identifying a potential threat action against an asset.

Risk assessment/analysis (qualitative, quantitative, hybrid)

Risk assessments (or risk analyses) are an important element of risk management. A risk assessment identifies an IT system that will be evaluated, including all in-place and planned security controls. It then attempts to identify the likelihood of various risks based on threats and vulnerabilities. Last, it attempts to prioritize the risks based on the potential loss to the organization.

True or false? A risk assessment is an ongoing process.

Answer: *False*. Risk management is an ongoing process, but a risk assessment is a point-in-time assessment of risk.

New threats and vulnerabilities might be discovered after a risk assessment is completed. If necessary, a new risk assessment can be completed in response to new risks or simply repeated periodically. Risk management practices can identify new risks and commission a new risk assessment when necessary.

EXAM TIP A risk assessment starts by identifying assets that will be evaluated. It then identifies threats and vulnerabilities with the goal of identifying the potential for threats to exploit vulnerabilities. Risks are then prioritized based on the potential loss to the organization.

True or false? A qualitative risk assessment is based on monetary data, such as the cost related to a risk.

Answer: *False*. Qualitative risk assessments are subjective and not based on numerical data or measurements.

The two primary types of risk assessments are known as quantitative risk assessments and qualitative risk assessments, which are defined as follows:

- Quantitative risk assessments depend on monetary values and other numerical data to identify the level of a risk. Historical data is used to identify the actual past costs and to extrapolate the potential future costs.
- Qualitative risk assessments are subjective and are based on the opinions of experts. Opinions are often gathered from surveys or during risk assessment meetings.

EXAM TIP Quantitative risks assessments use quantities such as monetary values related to a risk and gathered from historical data. Qualitative risk assessments use judgment calls based on opinions. A hybrid risk assessment uses a combination of both methods.

Risk assessments also consider the exposure factor from a risk. This is the proportion of an asset's value that is expected to be lost when a threat exploits a vulnerability. For example, an expert might determine that a fire will likely reduce the value of a building from US$100,000 to US$25,000. In this case, the exposure factor is 75 percent.

True or false? A qualitative risk assessment will often use a risk matrix providing a score based on the probability and impact.

Answer: *True*. Qualitative risk assessments are subjective and will often use the opinion of experts to identify the probability and impact of a risk.

One way a qualitative risk assessment is completed is via a survey that asks subject matter experts (SMEs) to rate risks based on a scale. Risks are often listed as threat/vulnerability pairs. The person conducting the assessment identifies a scale such as Low, Medium, or High and then asks participants to evaluate the probability (or likelihood) of each risk occurring, assuming the current countermeasures are in place. Further, the experts are asked to rate the impact of the risk if it occurs. Again, the person conducting the assessment identifies a scale, and for simplicity, it can use the same categories of Low, Medium, and High.

After the data is collected, the assessor can assign values to the words. For example, the probability values (sometimes called likelihood values) might be assigned the following values:

- Probability of risk occurring
  - Low = 10 percent
  - Medium = 50 percent
  - High = 90 percent
- Impact if risk occurs
  - Low = 10
  - Medium = 50
  - High = 90

After collecting and averaging the survey results, the risks might be calculated as follows:

- Risk 1 = Probability of 50 percent and Impact of 50 (Score = 12.5)
- Risk 2 = Probability of 90 percent and Impact of 50 (Score = 45)
- Risk 3 = Probability of 10 percent and Impact of 90 (Score = 9)
- Risk 4 = Probability of 50 percent and Impact of 10 (Score = 5)
- Risk 5 = Probability of 90 percent and Impact of 90 (Score = 81)

It's clear that the experts consider Risk 5 as the greatest risk.

NOTE Qualitative assessments are judgments. Even though the previous example converted the words low, medium, and high into numerical values, the assessment is still a qualitative assessment. The core decisions are made based on the opinions and judgment of the experts. Also, even though an item has a low score, management might decide to give it a high priority. For example, if there is a relatively low risk (10) of a fire but the impact is high (90), it has a low score of 9. However, management might choose to give it a higher priority.

Risk assignment/acceptance

Risk management includes multiple decisions, and a core decision is whether or not an organization wants to take steps to mitigate the risk or accept it. That decision is based on data collected during a risk assessment.

True or false? Purchasing insurance to reduce risk is an example of risk avoidance.

Answer: *False*. Purchasing insurance is an example of risk transference.

Risk management choices include the following:

- **Risk avoidance**   An organization can decide that the risk of a specific activity is too great and simply choose to avoid the risk by not participating in the risk activity. For example, a business might be thinking of expanding its business into e-commerce by hosting a website on its own server. After evaluating the options, it can decide the risks are too great and simply decide not to host an e-commerce website.

- **Risk transference**   The organization can also choose to transfer the risk to another entity. For example, the same business can choose to outsource the hosting of its website to a web hosting company. The web hosting company has assumed the risk related to protecting the server, and a contract can include penalties for noncompliance. Of course, a company that outsources services is still accountable to its customers for the ultimate service it provides. That is, the company still maintains a level of operational risk and reputational risk. Another common method of risk transference is purchasing insurance.

- **Risk mitigation**   The organization can also choose to mitigate or reduce the risk. This is done by implementing controls to reduce the vulnerabilities or the impact of the threat. For example, the business can ensure that input validation techniques are used for any programs that it develops for its e-commerce website. This reduces threats related to buffer overflows and SQL injection attacks.

- **Risk acceptance**   In some cases, the cost of a control outweighs the potential loss, so the risk is accepted. For example, a server hosting an e-commerce website could fail and this risk could be mitigated with a failover cluster to increase the server's availability. However, the cost of a failover cluster is high, so the business could choose to accept the risk that the server might fail and decide not to pay extra for a failover cluster.

Two other terms that are important to understand related to risk are *total risk* and *residual risk*, which are defined as follows:

- *Total risk* refers to the risk that is present without any security countermeasures. This is based on the asset value, threats, and vulnerabilities. Controls and countermeasures are implemented to reduce the threats and vulnerabilities, but they can't eliminate risk.

- *Residual risk* is the risk that remains after security controls are implemented. Management is responsible for risk management, including selecting controls to mitigate risks. Management is also responsible for any losses related to residual risk.

EXAM TIP Risks can be mitigated with controls or countermeasures, avoided by not participating in the risky activity, transferred through insurance or outsourcing, or accepted. Total risk includes all risk without any security controls, and residual risk is the risk that remains after security controls are implemented. Management is responsible for deciding which controls to implement and for any losses related to residual risk.

Countermeasure selection

Countermeasures reduce the risk by either reducing vulnerabilities or reducing the impact of a threat. However, an organization needs to carefully analyze a countermeasure to ensure that it meets its needs prior to selecting it.

True or false? When evaluating a countermeasure, the primary consideration is whether or not the countermeasure can eliminate or mitigate the risk.

Answer: *False*. When evaluating a countermeasure, an organization attempts to balance the risk against the financial concerns.

The primary purpose of any countermeasure is to reduce overall risk. Countermeasures reduce risk by addressing the vulnerability, the threat, or both. In some cases, it's possible to reduce or eliminate vulnerabilities with a countermeasure. Threats cannot be eliminated, but in some cases, it's possible to reduce the impact of threats.

Countermeasures are intended to reduce a risk but rarely eliminate it, and it's important to realize that it is impossible to eliminate all risk. One of the ways countermeasures are evaluated is by estimating their value based on a reduction of losses after they are eliminated. Three key terms when evaluating their value are as follows:

- **Single loss expectancy (SLE)** SLE is a monetary figure that identifies the actual loss for a single incident.
- **Annual rate of occurrence (ARO)** ARO identifies how many times the risk occurs in a year.
- **Annual loss expectancy (ALE)** ALE is calculated by multiplying the SLE by the ARO. For example, if one incident results in a US$1,000 loss (SLE) and the incident occurs four times a year (ARO), the ALE is US$4,000.

Effective countermeasures reduce the overall loss to the organization, typically by reducing the ARO. In some cases, the organization might reduce the SLE, but more often you'll see recommendations on countermeasure selection showing a reduction in the ARO.

True or false? A risk has an SLE of US$1,000 and an ARO of 20. A countermeasure is estimated to cost US$10,000 and will reduce the ARO by 5. This countermeasure should be purchased because it is beneficial to an organization.

Answer: *False*. The original ALE is US$20,000 (US$1,000 × 20). The savings is US$5,000 (ARO reduction of 5 × US$1,000). The control cost of US$10,000 is greater than the savings of US$5,000 and is not beneficial.

EXAM TIP Ensure that you understand SLE, ARO, and ALE and can calculate the ALE when given the SLE and ARO. You should also be able to determine whether the cost of a countermeasure is beneficial based on the reduction of the ARO. These are all used with a quantitative risk analysis.

Tangible and intangible asset valuation

Assets can have both tangible and intangible values. The tangible value is the actual cost of an asset, but the intangible value refers to other factors unrelated to the actual cost.

True or false? The loss of client goodwill is an example of a tangible value to a company.

Answer: *False*. The loss of client goodwill is an example of an intangible value.

Imagine that a company hosts an e-commerce website on its own server and the server fails. The cost to repair the server plus the lost revenue while it is being repaired is the tangible value. Tangible value includes the cost of hardware such as a computer system or network component, the cost of software applications, and the cost of data.

The intangible value includes other elements. Imagine that customers visit the website while it's down. Many might instead go to a competitor to get a similar product or service and never return, resulting in the loss of customers. Additionally, the negative publicity associated with the outage might discourage future potential customers from visiting. In this case, the intangibles include the loss of future revenue, the cost of regaining each customer who has left, and the customer influence with friends, family members, and/or business associates.

EXAM TIP Tangible assets refer to the cost of the asset. This can be the initial cost of the asset or the estimated cost to replace it. Customer goodwill is an example of an intangible asset. Using generally accepted accounting principles (GAAP), goodwill is often included on a company's balance sheet as an asset in addition to other net assets.

Can you answer these questions?

You can find the answers to these questions at the end of this chapter.

1. What are some threats related to employees?
2. A risk assessment compared risks against an email server. Based on the judgment of experts within the organization, one risk is determined to be a high risk to the organization but a second risk is determined to be a low risk. What type of risk assessment was completed?
3. What is residual risk?

4. A loss historically had a single loss expectancy of US$1,500 and an ARO of 10. A countermeasure costs US$3,000 and can reduce the ARO to 3. Based on this information, should the countermeasure be purchased?
5. An organization has determined that it will cost US$10,000 to replace an IT system. What type of asset valuation is this?

Objective 3.8: Manage personnel security

Personnel security starts before employees are hired by using effective candidate screenings. This helps an organization ensure that it knows whom it is hiring and also helps reduce its liability if the employee engages in fraudulent activities. During their employment, employees are provided information about what's required and expected of them. When their employment ends, the organization has the responsibility to ensure that the employee no longer has access to the organization's assets.

Exam need to know...

- Employment candidate screening (e.g., reference checks, education verification)
  For example: When should candidate screenings be performed? What is the purpose of candidate screenings?
- Employment agreements and policies
  For example: What is commonly included in an acceptable usage policy? When should an employee acknowledge the acceptable usage policy?
- Employee termination processes
  For example: What is a critical task related to user access when an employee leaves the organization? How much time should elapse between the time when an employee is fired and when access is revoked?
- Vendor, consultant, and contractor controls
  For example: What can be used to ensure that contractor accounts stop working when the contract ends?

Employment candidate screening (e.g., reference checks, education verification)

Before hiring new employees, it's important to complete screenings to verify their credentials. This includes procedures such as checking their references, performing credit checks, checking law enforcement records, and verifying their education claims.

True or false? An organization limits its liability by completing background checks during the candidate screening process.

Answer: *True.* An organization has a responsibility to perform due diligence when hiring new employees, and this includes performing relevant background checks.

Background checks are used to verify potential employees' credentials before they are hired. This helps the organization avoid liability. Imagine that a financial organization hires an individual convicted of financial fraud and puts this person into a position of trust to perform the same fraud. If the individual does indeed perform the same fraud, the organization could be taken to court for not exercising due diligence. Background checks also help protect sensitive information held by the organization. Individuals with lengthy criminal records are more likely to steal or reveal sensitive information than someone who has never done so.

Reviewing a resume and performing interviews provide some information about employee candidates. However, background checks provide a deeper level of understanding and can often uncover facts that a potential employee is trying to hide. Many employers also do drug testing in conjunction with their employee screening process.

EXAM TIP Background checks on potential employees help an organization avoid liability related to criminal behavior of employees. They also help an organization know who they have hired.

NOTE Although a background check does not necessarily investigate private information, it is becoming common for potential employers to perform Internet searches and check social media sites to identify an individual's social persona. An available white paper titled "The Ethics of Pre-Employment Screening Through the Use of the Internet" (*http://www.ethicapublishing.com/ethical/3CH4.pdf*) talks about some trends. Many job seekers know better than to use an email address like *IHateWork@outlook.com*. However, they still might use screen names that present them in an unfavorable light, and Internet searches might show this information. The ethics and legal implications of using these searches are debated by some people, but in general, information publicly available on the Internet is considered public information. Public information can be used for hiring decisions as long as the organization is not violating other laws, such as making decisions based on a person's race, color, religion, sex, or national origin.

MORE INFO Many organizations outsource background checks and potential employee screenings. For example, EmployeeScreenIQ (*http://www.employeescreen.com*) performs checks on personnel related to their credit, driving, and criminal records. They can also check references or perform full personal background checks when required.

Employment agreements and policies

It's common for an organization to require employees to agree to various employment agreements and policies. Employees are typically required to read and acknowledge these documents both when they are hired and periodically during their employment.

True or false? An employee policy will often include clear instructions to users on what is considered acceptable usage of an organization's assets.

Answer: *True.* An acceptable usage policy covers the rules and regulations related to using IT assets. It will also detail an employee's responsibilities.

Some elements that might be in an acceptable usage policy include the following:

- Instructions about acceptable and unacceptable use of email and the Internet.
- Statements making it clear that IT assets are the property of the organization.
- A privacy statement indicating what users should consider private communications when using the organization's assets. (Many organizations make it clear that users do not have any expectation of privacy when using IT equipment owned and maintained by the organization.)
- A statement identifying employee requirements to read and acknowledge the acceptable usage policy periodically.

EXAM TIP Employees are normally required to read and acknowledge an acceptable usage policy when they are first hired. They are then required to periodically review and acknowledge the policy, such as once a year as a part of a security training event. The usage policy can be signed either in person or electronically, based on the procedures used by the organization.

MORE INFO You can view a sample acceptable use policy on the SANS Institute website: *http://www.sans.org/security-resources/policies/Acceptable_Use_Policy.pdf.*

Employee termination processes

Most employees are loyal to the organization and will not take any malicious actions against it. However, when an employee is fired or terminated, it's not a wise decision to depend on the employee's loyalty. In the moments after an employee receives the shocking news that they no longer have a job, their actions aren't always predictable. As a precaution, the organization needs to take steps to prevent an ex-employee from taking malicious actions against the organization.

True or false? Employee accounts should be deleted as quickly as possible (within minutes) after an employee is terminated.

Answer: *False.* Employee accounts should be disabled during an exit interview, but they should not be deleted right away.

In some cases, administrators who had been terminated created another account with full administrative privileges within 10 minutes of leaving an exit interview and being told they were terminated. Even though the organization disabled their primary account sometime later, these administrators were able to use this second account to launch attacks against the organization remotely. Disabling all privileged access during the exit interview prevents this.

Accounts have unique identifiers used for access to resources, and they might be needed after the employee leaves. Because of this, the account should not be deleted right away. Instead, it is disabled, and after a supervisor or manager verifies that the account is no longer needed, it is deleted.

> **EXAM TIP** When employees are terminated, their access to all computer systems should be disabled as soon as possible. This prevents any malicious actions on the part of the ex-employee. Later, the account can be deleted when it is verified that it isn't needed.

> **MORE INFO** Chapter 1 included information about the access provisioning lifecycle and the importance of access reviews and audits. In addition to implementing policies related to termination processes, reviews and audits can be also used to verify that the policies are being followed.

Part of the termination process should also include steps to get employee-issued assets returned. For example, if an employee has been issued any type of hardware, such as a laptop or smartphone, this hardware needs to be returned.

Vendor, consultant, and contractor controls

Organizations frequently provide access to their assets to outside personnel for various purposes. Vendors sell products and services and are sometimes given limited access. Consultants are typically brought in for short periods of time for a specific project and need relevant access for that project. Similarly, contractors are hired for a specific contract period and need access to perform their job tasks during this time.

True or false? Setting account expiration is an effective control for contractor accounts.

Answer: *True*. Setting an account to expire when a contract expires is an effective security control.

Account expiration occurs automatically by disabling the account when the account expiration date arrives. It doesn't require an administrative interaction after the date is set to disable it and removes the possibility of human error. The account is still available if needed (such as if the contract is renewed) and can be re-enabled, granting the user all the same access.

Procedures should be in place to periodically review these temporary accounts to ensure that personnel do not have excessive privileges. That is, the principle of least privilege should be vigorously enforced.

> **EXAM TIP** Accounts created for vendors, consultants, and contractors are intended to be temporary. Account expiration controls should be used whenever possible to ensure that these accounts have limited lifetimes and are automatically disabled.

Can you answer these questions?

You can find the answers to these questions at the end of this chapter.

1. What elements should be included in an employee candidate screening?
2. How should employees acknowledge the contents of an acceptable usage policy?
3. When an employee is being terminated, who should be informed?

Objective 3.9: Develop and manage security education, training, and awareness

Personnel cannot be expected know what an organization's security goals are unless they are communicated to them. They might assume that the current organization has the same goals as their previous organization. If the previous organization had no security program, personnel might assume that this organization is the same. All of this is countered with an effective security education, training, and awareness program.

Exam need to know...

- Security education, training, and awareness
  For example: What is the goal of education, training, and awareness programs? Who is the target audience of these programs?

Security education, training, and awareness

Security education, training, and awareness programs help all personnel understand an organization's security practices. They help users understand the underlying risks and the value to themselves and the organization when security practices are followed.

True or false? Effective security education, training, and awareness programs are focused primarily on regular users performing day-to-day tasks.

Answer: *False*. Effective security education, training, and awareness programs need to reach all personnel within the organization. This includes regular users, but it also includes management personnel, administrative personnel, technicians, and security personnel.

Security education, training, and awareness programs are both generic and targeted. Some elements are designed for all personnel, such as notification of a recently released virus or phishing email. Other programs are targeted to meet specific audiences. For example, administrators are trained on how to implement and maintain security devices such as firewalls and intrusion detection/prevention systems.

True or false? A primary purpose of security education is ensuring that personnel can be held accountable for their actions.

Answer: *False.* A primary purpose of security education is to help users understand their role and responsibilities related to security with an overall goal of reducing security incidents.

Personnel can be held accountable for not following established security practices, but that isn't a primary purpose of security education. However, it is common to have personnel read and acknowledge an acceptable use policy in conjunction with periodic training.

> **EXAM TIP** Security education, training, and awareness programs are used to help all personnel understand the value of security. These programs help regular users understand common attacks used by malicious individuals, such as social engineering attacks. The programs also help ensure that all personnel recognize their responsibility to exercise basic security practices to protect their resources and the organization's assets.

True or false? Security education, training, and awareness programs often provide the reasoning behind security rules.

Answer: *True.* When personnel understand the reasoning behind a rule, they are often more apt to follow it.

For example, telling users not to display graphics when they receive an email from an unknown sender might be a rule they choose to ignore. On the other hand, you could explain that graphics are held on Internet servers and often include beacons that verify their email address. If users display the graphics, the beacon verifies their email address as valid and they can count on receiving more spam. When they understand this, they are more likely to understand and abide by the rule.

Can you answer these questions?

You can find the answers to these questions at the end of this chapter.

1. What is the goal of an organization's security education and training program?
2. Who is the target audience of security education and training programs?
3. In addition to explaining rules, what is also provided in training related to the rules?

Objective 3.10: Manage the security function

This objective focuses on many of the underlying managerial issues related to security programs. Management needs to ensure that security has an adequate budget and other resources to meet the security goals. Metrics can be used to validate the security function on a day-to-day basis. Additionally, it's possible to perform regular assessments to evaluate the completeness and effectiveness of a security program.

Exam need to know...

- Budget
  For example: How can the projected benefits of a countermeasure be calculated?
- Metrics
  For example: What can be measured to evaluate security? What are some characteristics of valid metrics?
- Resources
  For example: In addition to money, what are some other resources needed to support security?
- Develop and implement information security strategies
  For example: What primary security document is used to reflect an organization's security strategies?
- Assess the completeness and effectiveness of the security program
  For example: How can an organization determine whether personnel are susceptible to social engineering attacks?

Budget

In many companies, the business element is devoted to generating revenue. In contrast, IT support and security functions do not generate revenue but instead are considered cost centers that take away from the revenue. Budgets help ensure that the IT and security functions are supporting the primary mission of the organization while not taking too much money away from the revenue.

True or false? A cost-benefit analysis can be used to evaluate a countermeasure before it is purchased.

Answer: *True*. A cost-benefit analysis identifies the potential value of a security countermeasure against its cost.

Some of the terms related to a countermeasure used within a cost benefit analysis include the following:

- **Loss before countermeasure** This is the ALE without the countermeasure.
- **Loss after countermeasure** This is the predicted ALE after the countermeasure is implemented. Countermeasures rarely eliminate the risk but instead reduce the ARO, resulting in a lower ALE.
- **Projected benefits** This is calculated as the loss before the countermeasure minus the loss after the countermeasure. Because SLE, ALE, and ARO are all calculated on an annual basis, this provides the projected benefits for the year.
- **Cost of countermeasure** This includes the cost to purchase, implement, and maintain the countermeasure. If personnel need to be trained on how to implement, maintain, or use the countermeasure, the training costs should also be included.

EXAM TIP If the projected benefits of a countermeasure are greater than the cost of the countermeasure, the countermeasure is beneficial to the organization. If the benefits are less than the cost, the countermeasure should not be purchased. If the benefits and the cost of the countermeasure are close to equal, a return on investment (ROI) study can be performed to compare the long-term benefits and costs.

Metrics

Metrics allow an organization to measure the effectiveness of a process or system. Within IT security, metrics can be used to identify the effectiveness of the security function.

True or false? The number of security incidents for a given time period can be used as a metric.

Answer: *True.* The number of security incidents (monthly, quarterly, and/or annually) identifies the effectiveness of the organization's overall security function.

By comparing current metrics against historical metrics, an organization can identify differences in the security function. They can also identify trends in security issues and diagnose problems. Beyond just looking at the total number of security incidents, an organization can also measure the following items:

- **Malicious software (malware) infections** This can measure the total number of infections, but it's also important to measure the method of infection. For example, malware might be delivered through email, when users visit websites, or from USB flash drives. Different controls can be used to mitigate the threat based on how it is getting into the organization.
- **Intrusion detection and prevention system reports** Both false positives and true positives should be examined. Often, the threshold of these systems can be tweaked to reduce the number of false positives without sacrificing the ability to detect true positives.
- **Spam** The amount of filtered spam can be regularly checked to verify that it is effective. Filters should block as much spam as possible without blocking valid email, so a spam filter will rarely filter out 100 percent of spam.
- **Results from vulnerability scans** Periodic vulnerability scans can be compared against each other to identify trends and validate other security functions. For example, by checking systems for current updates, a vulnerability scan validates a patch management program. If systems are consistently found to be out of date, the organization can take a closer look at the patch management program to identify what can be changed to improve it.
- **Outages** Any outage that results in a loss of availability should be documented. The organization can identify an acceptable level of outages for any system and then measure the actual outages to determine its performance.
- **Backups** Personnel can periodically perform a test restore of data from a random backup. For example, a weekly test restore allows an organization to validate the effectiveness of the backup system. Ideally, it is always possible to restore data from backups, but in actual practice, many variables affect the quality of backups.

True or false? Metrics should be consistently measured and provide actionable data.

Answer: *True.* Metrics must be measured the same way to ensure that the comparisons are meaningful, and the data needs to give decision-makers information they can use.

For example, comparing an outage time from a noncritical system against an outage time for a critical system doesn't provide much meaning. In contrast, comparing outages for a critical system before a countermeasure was added against outages for the same system after adding a countermeasure provides information about the value of the countermeasure.

> **EXAM TIP** Metrics provide actionable information that can be used by managers. The actual metrics used to measure security can vary between organizations, based on their goals. An important consideration when choosing metrics is to ensure that they are measured consistently.

> **MORE INFO** The SANS Institute InfoSec Reading Room has a white paper titled "Gathering Security Metrics and Reaping the Rewards." You can access the paper here: *http://www.sans.org/reading_room/whitepapers/leadership/gathering-security -metrics-reaping-rewards_33234.*

Resources

Resources within an organization include money, people, hardware, and software. There isn't a specific amount of resources needed. Instead, any organization needs to have sufficient resources to meet its security goals.

True or false? It's important for every organization to have dedicated security staff to address security issues.

Answer: *False.* Large organizations will often have dedicated staff, but this isn't feasible for every organization.

Many smaller organizations have personnel who share security duties. The amount of resources dedicated to security varies in different organizations and is dependent on the goals and mission of the organization.

In some reactive organizations, security goes through feast and famine cycles. Security resources are minimized until an incident occurs. After an incident, security resources are increased dramatically and then gradually removed until another incident.

True or false? The cost of backup tapes is considered a security resource.

Answer: *True.* Backups are created to increase availability by ensuring that backups are available if the primary data is lost.

The cost of backups is often underestimated until an organization suffers a substantial loss of data. When data is lost but easily restored, management often doesn't recognize the value of the backups because the process is transparent at the managerial level. In some situations, the value of the backup process is not recognized. On the other hand, when data is lost but backups are not available, it's a crisis situation.

> **EXAM TIP** An organization needs enough resources to meet its security needs. This varies and is based on the organization's goals and mission.

Develop and implement information security strategies

An organization develops and implements information security strategies to support its overall security function. Elements such as a security steering committee can be used to ensure that security goals are identified.

True or false? Security strategies are reflected in an organization's security policy.

Answer: *True*. Security policies are used to reflect the organization's security goals, along with the overall mission and vision of the organization.

Assess the completeness and effectiveness of the security program

Audits and reviews are regularly done to assess the effectiveness of a security program. These can be done internally, by external experts, or as a combination of both.

True or false? Fake phishing emails sent to employees can be used to test employee understanding of email risks.

Answer: *True*. An organization can verify employee understanding and compliance with security policies by sending fake emails and measuring their response.

Some companies, such as Knowbe4.com, provide a variety of online training and also provide security assessment services. For example, they can send a fake phishing email with an attachment, graphic link, or website link to employees and provide a report back to the organization identifying personnel who responded with unsafe practices. In real life, these unsafe practices can result in malware being installed on user systems or users revealing sensitive information. In an assessment test, there aren't any risks but it does identify potential problems with the security awareness program.

When an organization understands risky activity taken by personnel, it has a better chance of developing methods to change the risky behavior. It could be that personnel aren't receiving the proper training, that the training isn't effective, that personnel don't believe the organization takes security seriously, or something else.

> **EXAM TIP** Regular audits and reviews are an integral element of assessing the effectiveness of a security program. They should examine the records and logs created during day-to-day activities to identify how the security program is implemented.

True or false? Penetration tests can check an organization's operational security procedures.

Answer: *True*. Penetration tests can check operational security procedures, such as if personnel are willing to give out data from a social engineering attack.

Penetration tests attempt to exploit the physical, operational, or electronic elements of an organization. Physical tests check the physical perimeter of an organization, such as if unauthorized individuals can gain access through tailgating or some other method. Operational tests help determine whether personnel are susceptible to social engineering tactics. Electronic tests check security of IT resources, including networks and servers.

> **EXAM TIP** Penetration tests should be performed only after receiving explicit permission from senior management. External penetration testers should perform a test only after receiving written permission.

> **MORE INFO** Chapter 7 mentions penetration tests (including white box, black box, and gray box testing) in the context of vulnerability management in Objective 7.5. When approved, an electronic penetration test often starts with a vulnerability assessment to detect weaknesses. It then determines whether these weaknesses can be exploited.

Can you answer these questions?

You can find the answers to these questions at the end of this chapter.

1. A cost-benefit analysis was recently completed and provided the following information. A countermeasure has a projected cost of US$5,000. If it is implemented, the projected losses are expected to be US$5,000. If it is not implemented, the projected losses are expected to be US$15,000. Should this countermeasure be purchased?
2. What can be measured to determine the effectiveness of antivirus software?
3. What is the minimum number of resources that should be dedicated to an organization's security function?
4. What outlines an organization's strategic goals related to IT security?
5. What is the purpose of a penetration test?

Answers

This section contains the answers to the "Can you answer these questions?" sections in this chapter.

Objective 3.1: Understand and align security function to goals, mission, and objectives of the organization

1. The primary goal of any IT security program is to protect against the loss of confidentiality, integrity, and availability of an organization's assets. The assets will often be different and the methods used to protect them might also be different, but the goal is to protect against losses.
2. Management should review and approve the security policy. It's important that the security policy includes information related to the organization's goals, mission, and objectives. When approving the security policy, management provides direction to others within the organization to develop and implement controls to support the organization's vision.

Objective 3.2: Understand and apply security governance

1. A security-based steering committee makes decisions related to the overall direction of a security program.
2. The example data is PII. PII includes any information that can be used to distinguish an individual, including their name, its mother's maiden name, and much more.
3. The following five principles are in COBIT version 5:
    a. Meeting stakeholder needs
    b. Covering the enterprise end-to-end
    c. Applying a single, integrated framework
    d. Enabling a holistic approach
    e. Separating governance from management
4. An organization has a responsibility to identify risks that might compromise confidentiality, integrity, or availability of its assets as a part of due diligence.
5. An organization has a responsibility to evaluate risks as a part of due care. An organization isn't expected to eliminate all risks, but it is expected to use basic risk management practices to evaluate the risks and make reasonable decisions about how to address these risks.

Objective 3.3: Understand and apply concepts of confidentiality, integrity, and availability

1. The two primary methods of preventing the loss of confidentiality are encryption and effective access controls. Both methods prevent unauthorized individuals from accessing data.
2. Hashing methods can detect modifications to any file, including system files. A hash is first created with a known good version of the file. Later, another hash is taken of the file and compared to the original. If the hash is different, it verifies that the file has been modified and lost integrity.

3. RAID, UPS, and failover clusters are all designed to increase the availability of systems. They increase availability in different ways, but they all try to eliminate single points of failure.
4. The organization has lost confidentiality and availability. Confidentiality is lost because the thief can now access this data. Availability is lost because the organization is no longer able to restore the data by using these tapes. If the organization had created two sets of backup tapes (one for on-site and one for off-site), it presents less of an impact on availability.

Objective 3.4: Develop and implement security policy

1. External laws and regulations must be considered when creating a security policy. The security policy reflects the overall security vision of the organization and provides direction from executive management. When creating or approving the security policy, executive management has a responsibility to abide by relevant laws and regulations.
2. A security baseline provides a starting point for security. Security can then be enhanced beyond the baseline.
3. Documents that show how to perform tasks such as backups and restores are procedures.
4. The security policy should be reviewed regularly, and only executive management can approve changes to the security policy.

Objective 3.5: Manage the information lifecycle (e.g., classification, categorization, and ownership)

1. Senior management has a responsibility to ensure that data classifications have been defined and are understood by personnel in the organization.
2. Data owners are responsible for ensuring that their data is properly marked, labeled, and protected.
3. Data classifications help prevent the loss of confidentiality of valuable data. Personnel recognize that data with higher classification levels is more valuable and take extra measures to protect it.

Objective 3.6: Manage third-party governance (e.g., on-site assessment, document exchange and review, process/policy review)

1. IT governance refers to the processes in-house used to ensure that IT systems coincide with the vision of the organization. Third-party governance refers to the external rules and regulations that apply to the organization.
2. HIPAA applies to any organization in the United States that handles any type of health information. This includes any information about past, present, or future physical health; mental health; physical condition; and payments for medical services.

3. Any publicly traded company that provides reports to the United States Securities and Exchange Commission must abide by the Sarbanes-Oxley Act.

Objective 3.7: Understand and apply risk management concepts

1. Employees can cause both accidental and malicious actions against an organization's assets. Accidental actions include unauthorized access to data (innocently), outages after making a change (when a change management program isn't in place), and damage to hardware (such as by accidentally spilling liquid). Malicious actions include theft of data, sabotage, fraud, and theft. Various security controls help to mitigate all of these threats.
2. A risk assessment that is based on the subjective opinion of experts is a qualitative risk assessment. Qualitative risk assessments typically attempt to identify the probability and impact of risks as a method of prioritizing them.
3. Residual risk is the risk that remains after security countermeasures are implemented. In contrast, total risk is the risk that is present without any countermeasures.
4. Yes, the countermeasure should be purchased. Without the countermeasure, the ALE is US$15,000 (US$1,500 × 10). With the control, the ALE is US$4,500 (US$1,500 × 3). This provides a cost benefit of US$10,500 (US$15,000 - US$4500). The cost of the control (US$3,000) is significantly less than the benefits of the control.
5. The actual or estimated cost to replace an IT system is an example of a tangible asset.

Objective 3.8: Manage personnel security

1. Employees should be prescreened based on the responsibilities of their job. This often includes checking references, credit records, criminal records, and education claims. Many organizations also include drug testing.
2. Employees should be required to read and acknowledge an acceptable usage policy when they are hired and periodically afterwards. This can be done with a physical signature or with an electronic signature. For example, after performing online training, an employee could be required to read the policy, check a box indicating that they have read it, and acknowledge the contents. They then click another button, such as OK. The user is identified based on the logged-on account, and their actions are logged.
3. When an employee is terminated, someone having the ability to disable the employee's account should be informed. Who this person is depends on internal procedures and could be someone in the human resources department. However, the goal is to ensure that the account is terminated before the employee leaves the exit interview.

Objective 3.9: Develop and manage security education, training, and awareness

1. The goal of any security, education, and training program is to reduce security incidents. When people understand the security requirements and the benefit of following the requirements, they are more apt to follow them, resulting in fewer security incidents.
2. Training is for all personnel. Some training might be generic and designed for everyone whereas other training is designed for specific groups. However, all personnel should receive some type of training.
3. Personnel are often provided information explaining the reasoning behind the rules. When personnel understand the risks and how the rules help them and the organization, they are more likely to abide by the rules.

Objective 3.10: Manage the Security Function

1. The cost-benefit analysis indicates that the countermeasure should be purchased. The projected benefits are calculated as loss before countermeasure minus loss after countermeasure. This is US$15,000 – US$5,000 or US$10,000. The countermeasure is US$5,000. The benefits of the countermeasure (US$10,000) are significantly greater than the cost of the countermeasure (US$5,000).
2. Metrics that show the number of malware infections detected by antivirus software can provide an effective measure. These numbers should be compared to determine where the malware was discovered, such as at an email server, at the boundary to the Internet before it reached a user's computer, or on the user's workstation.
3. The minimum number of resources needed for an organization's security function ensures that the security goals are met. The minimum varies and is dependent on the organization.
4. Security policies outline the strategic goals of an organization. These are often created with input from a security steering committee in a large organization.
5. The primary purpose of a penetration test is to determine whether vulnerabilities can be exploited. Vulnerability assessments identify weaknesses, and penetration tests verify that an attacker can exploit a weakness.

CHAPTER 4

Software development security

This domain of the CISSP exam focuses on the risks associated with software development and the steps an organization can take to minimize these risks. There is a heavy focus on the software development lifecycle (SDLC) and the system development lifecycle, and you should have a general idea of the software development processes. However, you don't need to know the details of individual programming languages because the majority of the security issues are similar between languages. Similarly, the overall security goals are the same regardless of the language used—applications should maintain the integrity of data and the application and prevent loss of availability.

This chapter covers the following objectives:

- Objective 4.1: Understand and apply security in the software development lifecycle
- Objective 4.2: Understand the environment and security controls
- Objective 4.3: Assess the effectiveness of software security

Objective 4.1: Understand and apply security in the software development lifecycle

At one time, software developers wrote an idea on a lunch napkin and started coding the application that afternoon. Today, applications are much more complex and require structured processes to ensure that the progress can be monitored and controlled. Many different models are available to monitor the process throughout the lifetime of any application. Most are based on the system development lifecycle or the software development lifecycle. Organizations that use these processes can be rated with a capability maturity model. Ratings identify the level of maturity of an organization's processes and its ability to repeat its successes. All organizations need to be aggressive with a change management program to prevent disruptions or outages caused by unauthorized changes.

Exam need to know...

- Development lifecycle
  For example: What are common stages identified in a system development lifecycle? What is the waterfall model?
- Maturity models
  For example: How many levels are in the Capability Maturity Model? How is a maturity level assigned?
- Change management
  For example: What is the primary purpose of change management? What types of changes should be included in change management procedures?

Development lifecycle

The software development lifecycle (SDLC) separates software development activities into different phases. Multiple SDLC models are used, and many identify the phases differently, but there are similarities between the models.

> **EXAM TIP** The software development lifecycle is considered a subset of the system development lifecycle, and you might see both identified with the SDLC acronym in different study materials. Even though the objectives refer specifically to the software development lifecycle, you can also expect to see questions on the system development lifecycle.

True or false? Security controls are first identified in the implementation stage of the software development lifecycle.

Answer: *False*. Security is included throughout the software development lifecycle, and specific security controls should first be considered during the initiation, development, or design stage.

Some models use more stages by dividing tasks within a stage. For example, instead of combining initiation, system concept development, and planning into a single stage, another model might divide these tasks into three separate stages. However, the following stages provide an overview of the concepts included in the different models:

- **Initiation, system concept development, and planning** The scope of the system is identified here, and the system is evaluated for cost benefits and feasibility. Security considerations are included at this stage.
- **Requirements analysis and design** During this phase, developers design the application while ensuring that security controls are included from the start. Methods of validating, verifying, and testing the plan are identified along with a review of the design. A prototype test might be conducted during this phase to ensure that the concept is possible. Prototype tests are also used to validate the security capabilities.
- **Development, integration and test** This includes all types of testing, including individual unit or module testing and integration testing with other systems.

- **Implementation, operations, and maintenance** During this stage, the system is operating as intended. Key security actions during this stage are change management and configuration management to ensure that the system stays in a secure state. Continuous monitoring and risk analysis techniques are used to detect any issues that might arise. Routine security auditing is performed during the operations of a system or application to monitor user activity, and results can be used to hold users accountable for unauthorized activity.
- **Disposition/disposal** When the system reaches the end of its lifecycle, a transition plan is implemented to dispose of hardware and software. A security concern at this stage is ensuring that media is sanitized and does not include any data remnants.

MORE INFO NIST SP 800-64, "Security Considerations in the System Development Life Cycle," provides good explanations of the different phases of the system development lifecycle with a focus on security. You can download any SP 800 documents here: *http://csrc.nist.gov/publications/PubsSPs.html*. Also, NIST has published a System Development Life Cycle brochure that lists key NIST documents for each phase of the system development lifecycle. You can download the brochure here: *http://csrc.nist.gov/groups/SMA/sdlc/documents/SDLC_brochure_Aug04.pdf*.

Many different software development lifecycle models are available, including the following:

- **Waterfall** In this model, progress flows from the top down as in a waterfall. The phases are: requirements, design, implementation, verification, and maintenance.
- **Spiral** The spiral model attempts to combine top-down and bottom-up concepts. It repeats four phases to progressively build the application from the bottom up and also allows viewing of the final product as it is built and improved.
- **Prototyping** This model uses prototypes to test a concept. When the concept is validated, the prototype is enhanced to create a final product.
- **Rapid application development (RAD)** This model minimizes advanced planning and instead uses rapidly developed prototypes. The prototypes are then designed and modified in an iterative process until the product meets the customer's needs.
- **Incremental** In this model, the application is built and tested in increments, where each increment adds one or more additional features to the application.

EXAM TIP Security must be considered at each stage of the software development lifecycle, and because of this, many organizations consider it essential that software developers understand the development lifecycle. If security is addressed only at the end of the process, it is often inadequate.

Maturity models

Maturity models are used to define the level of maturity of a software development company and its ability to develop software applications in a predictable manner. A common model used by many companies is the Capability Maturity Model (CMM).

True or false? The Capability Maturity Model Integration (CMMI) includes five maturity levels.

Answer: *True*. These levels define the level of consistency that an organization applies to its design, development, and testing of software products and services.

The five CMMI levels are as follows:

- **Level 1: Initial**   At this level, processes are performed on an as-needed basis and are usually chaotic. The organization can produce products and services, but they often exceed budget and schedule milestones. Additionally, they are not necessarily able to repeat their successes.
- **Level 2: Managed**   Software development projects are planned in accordance with an organization's policies. Projects are monitored, controlled, and reviewed.
- **Level 3: Defined**   The organization has established standards, procedures, tools, and methods for software development. These are understood by personnel in the organization and improved over time.
- **Level 4: Quantitatively Managed**   The organization has established quantitative objectives for quality and uses these to measure the performance of projects.
- **Level 5: Optimizing**   This is the highest CMMI level. An organization at this level uses a continuous improvement process to improve existing processes and procedures.

Organizations are appraised by using specific guidelines, and these appraisals identify the organizations' current level of compliance with a CMMI level. Organizations can then implement action steps to achieve compliance at a higher compliance level. If an organization has not gone through an appraisal process, it is considered to be at Level 0.

> **EXAM TIP**   The five levels of the CMMI identify best practices that organizations can use to define, track, and improve their processes. The CMMI levels are very useful for outside organizations when purchasing software products and services. An outside organization can use these levels to identify vendors that are considered reliable and trustworthy because organizations at higher levels are less likely to go over budget and more likely to meet scheduling milestones.

> **MORE INFO**   You can download the CMMI for Development, Version 1.3, published by the Software Engineering Institute at Carnegie Mellon here: *http://www.sei.cmu.edu/reports/10tr033.pdf*. Chapter 3, "Information security governance & risk management," describes the five levels.

Operation and maintenance

Operation and maintenance of software is an important phase within the software development lifecycle. During this time, users are actively using the software.

True or false? Software developers are not involved with software development during the operation and maintenance phase.

Answer: *False.* Software developers still need to be available to evaluate software bugs and create updates when necessary.

Software is rarely released completely bug-free. As bugs are discovered, software developers create and test updates that are then released to the customers. This includes software that is developed for in-house users and software that is developed to be sold to outside organizations.

> **EXAM TIP** When budgeting for software, the organization needs to consider ongoing maintenance. When software is developed in-house, the organization needs to retain software developers for the lifetime of the software. Additionally, all code must be backed up and protected. If a bug is discovered a year after the software is released, the developers need to have access to the original code to create an update.

Change management

Change management practices are designed to control changes and prevent unauthorized changes from causing unexpected outages. Systems and software are often complex, and a change to one element can have unintended negative side effects on another system. A change management system ensures that changes are analyzed prior to implementing them.

Within the context of software development, unauthorized changes often impact other elements of the application. For example, a change in one module might negatively affect multiple other processes that use this module.

Additionally, a change management system ensures that changes are tested when necessary and that these changes are documented. Testing ensures that potential side effects from a change are identified before a change is implemented, and documentation ensures that all personnel are aware of the changes.

True or false? Significant changes that are controlled with change management practices include changes in budget constraints.

Answer: *False.* A budgetary constraint does not need to go through change management. However, change management documentation does assist with budget decisions by helping to identify out-of-scope expenses.

Some common changes that should go through a change management process include the following:

- **Changes to software** This includes upgrades, patches, and configuration changes to software applications and operating systems.
- **Changes to hardware** Any additions or modifications to the hardware should be evaluated with a change management process.

- **Change in the location** Moving a system to a different room, building, or geographical location should be reviewed. It can impact the electrical, heating, ventilation, and air conditioning (HVAC) requirements. Moving a system can also impact its network availability.
- **Change in data value** If a system begins to process data that is considered more sensitive or of a higher value, the system will need additional security controls. Before this change, the system should be evaluated to ensure that the security controls are in place.

EXAM TIP Unauthorized changes frequently cause unexpected outages and directly impact availability of systems. Personnel with the privileges to make changes should have a clear understanding of change management procedures and be held accountable when these procedures are not followed. Change management includes the documentation of changes, which tracks all changes that have been authorized and implemented.

Can you answer these questions?

You can find the answers to these questions at the end of the chapter.

1. Name two models that are commonly used for software development.
2. Who benefits from a CMMI rating?
3. What is an unintended side effect when change management procedures are not used?

Objective 4.2: Understand the environment and security controls

For this exam objective, you should have a good understanding of the common security issues associated with different programming languages and the ways that they can be mitigated. Compiled applications can be reverse engineered, but some security procedures, such as encryption and code obfuscation, reduce the success of attackers. Many attacks, such as buffer overflow attacks and backdoor attacks, can be prevented with the use of security principles such as input validation techniques and completing a code review prior to releasing the application. Configuration management practices ensure that applications are deployed in a consistent state and, combined with change management practices, ensure that applications remain in a secure state.

Exam need to know...

- Security of the software environment
  For example: Should an application provide verbose errors to users when an error occurs? How should you protect a password that must be embedded in an application's code?

- Security issues of programming languages
  For example: What is mobile code? How can a developer validate the integrity of code after it has been released?
- Security issues in source code (for example, buffer overflow, escalation of privilege, backdoor, and so on)
  For example: What causes a stack overflow? What is a backdoor?
- Configuration management
  For example: At what stage of the SDLC should configuration management occur?

Security of the software environment

Security of the software environment refers to the different ways that software interacts with users, other applications, and network services. It also includes methods used to ensure the validity and integrity of data.

True or false? The Generic Security Service Application Programming Interface (GSS-API) defines standards for security services used in applications.

Answer: *True.* GSS-API is defined in RFC 2743 (*http://tools.ietf.org/html/rfc2743*). It provides guidance about how to create standardized programming interfaces for authentication, access control, monitoring, and logging.

> **MORE INFO** While the GSS-API is generic, some other documents provide guidance for specific languages. For example, RFC 2744 (*http://tools.ietf.org/html/rfc2744*) identifies specific interfaces for the C programming language, and JSR-000072 (*http://jcp.org/aboutJava/communityprocess/review/jsr072*) provides specific guidance for the Java programming language. Java is a single language that can be executed on almost any operating system by using an operating system–specific Java Virtual Machine (JVM). The JVM converts the applet into machine-level code for the specific system.

True or false? Error messages should provide as much detail to users as possible.

Answer: *False.* Detailed error messages provide information that can be used by an attacker. Instead, error messages in release versions of applications should be kept basic without providing information about the system architecture. Developers can use verbose error messages during development, testing, and debugging phases.

> **EXAM TIP** Detailed error messages can give attackers information about the operating system version, installed service packs, and installed updates. They can also give detailed information about the applications, such as the version of web server software or the version of database software. An educated attacker can easily look at a SQL error message and determine whether it is a Microsoft SQL Server database, an Oracle database, a MySQL database, or something else.

One method of ensuring the validity and integrity of data is by using time and date stamps. For example, a payroll application can use time and date stamps to check data before using it. Only entries within a certain time period will be used to calculate a paycheck.

Validity of data can also be checked based on the data's origin. For example, personnel might be restricted from clocking in or out from any location other than an on-site location. Many Internet-based payroll programs are available that allow personnel to access their payroll information from anywhere while restricting their ability to clock in or out to a specific domain or IP address.

> **EXAM TIP** Time and date stamps are commonly used to prevent many types of replay attacks. For example, Kerberos uses time stamps along with Advanced Encryption Standard (AES) symmetric encryption to prevent someone from intercepting a ticket and using it in a replay attack.

True or false? If a software application needs a password, the password must be compiled so that it cannot be read.

Answer: *False*. Passwords should be stored in a secure hash format. Tools are available to reverse engineer compiled code, and these tools might be able to discover the password if it is only compiled and not encrypted or hashed.

> **MORE INFO** A hash is simply a fixed-size string of characters created from a password, message, or file. As long as the original data is the same, the hashing algorithm will always create the same hash. Hashing and other cryptography topics are covered in Chapter 5, "Cryptography."

> **EXAM TIP** Only secure hashing methods should be used to store passwords within an application. For example, Secure Hash Algorithm 256 (SHA-256) is a strong hashing algorithm, but Message Digest 5 (MD-5) is susceptible to collisions and should not be used for this purpose.

Another protection against reverse engineering is code obfuscation. Developers use easily recognizable names for variables and modules when developing the code. For example, a password variable might be named strPassword (indicating a character string for a password) and a module to check authentication might be named ValidateCredentials.

When the code is compiled with code obfuscation, the strPassword variable might be renamed as x4 and the ValidateCredentials module might be renamed as e8. This obscures the purpose of the variables and modules, making it more difficult to reverse engineer the code.

True or false? If an application suffers a failure from a denial of service (DoS) attack, it should be designed to fail in such a way that all access is denied.

Answer: *True*. Failures from any type of attack, including a DoS attack, should fail in a safe or secure manner. This is also known as failing with a "deny all" philosophy.

> **MORE INFO** The Open Web Application Security Project discusses the dangers of detailed error messages and the importance of failing in a safe manner on their Error Handling, Auditing, and Logging page: *https://www.owasp.org/index.php/Error_Handling,_Auditing_and_Logging*. The Resource Exhaustion page also discusses failing in a safe manner: *https://www.owasp.org/index.php/Resource_exhaustion*.

It's common for applications to access data in back-end databases, data marts, or data warehouses. Most online systems will use a basic database, while data mining systems use combined data marts or data warehouses.

True or false? An online transaction database (OLTP) should meet at least three elements of the ACID model.

Answer: *False.* A core requirement for any database, especially an OLTP database, is to meet all four elements of the ACID model, commonly called the ACID test.

> **NOTE** ACID is an acronym for atomicity, consistency, isolation, and durability. *Atomicity* refers to how transactions are committed to a database only if they complete as a whole or are rolled back if any element of the transaction doesn't complete. *Consistency* refers to how committed data is consistent with other data in the database. *Isolation* refers to how database transactions complete without interacting with other transactions. *Durability* refers to how a database can survive a failure and still ensure that committed transactions are applied and that uncommitted transactions are rolled back.

> **EXAM TIP** Databases use transaction logs to track transactions and checkpoints to commit transactions to the database. A checkpoint process periodically examines the transaction log to identify committed transactions. Committed transactions are applied to the database, and this action is recorded in a checkpoint log. If a database fails due to a power failure or any other problem, the database system can examine the transaction and checkpoint logs, commit any transactions that haven't been committed yet, and roll back any partial or uncommitted transactions.

True or false? A primary key in a database table uniquely identifies a row.

Answer: *True.* A primary key is a column in a table, and in each row (also referred to as a tuple or a record), the primary key is unique. A relational database uses the primary key to identify the row and can relate two tables by using the primary key in one table and a foreign key in another table.

> **EXAM TIP** Normalization is an important process used to optimize databases. Most OLTP databases are normalized to third normal form (3NF). Concurrency controls are used to prevent processes from modifying data while it is being read or modified by another process. If concurrency isn't managed, it can result in a deadlock situation where both processes lock the data, each waiting for the other process to release it. A deadlock can significantly impact the performance of the application using the database.

True or false? Internet-based applications and services exchange information by using SOAP.

Answer: *True.* SOAP is an XML-based language used to exchange information between systems over the Internet. It was developed as an alternative to Remote Procedure Calls (RPCs). SOAP was originally an acronym for Simple Object Access Protocol.

Security issues of programming languages

Generally, security issues are similar for any programming languages. An exception is mobile code, which is commonly used on websites. Websites are vulnerable to a variety of attacks and must be checked for different vulnerabilities.

True or false? Mobile code is any code that can be transferred across a network and run on another system without being installed.

Answer: *True*. Mobile code includes a variety of different types of scripting languages, such as VBScript and JavaScript. It also includes mini-applications such as ActiveX controls, macros, Java applets, and Flash animations.

Mobile code is often embedded within a webpage and is executed on a user's system without any additional interaction. In some cases, this allows the mobile code to execute outside of a web browser, which presents risks to users.

Cross-site scripting (XSS) is where an attacker can inject malicious code into a website. When users visit the website, the malicious code is downloaded and executed on their system. An XSS attack doesn't directly affect an organization's website availability, but it can cause significant public relations problems. If users recognize that they are being infected after visiting a legitimate website, they might simply stop going to the site.

> **EXAM TIP** In XSS attacks, attackers inject client-side scripts into webpages on the web server. Because it is a client-side script, it runs on user systems that visit the website rather than on the web server. XSS scripts often exploit vulnerabilities in the document object model (DOM) used by web browsers. Successful attacks allow attackers to obtain cookies and other information from user systems.

Drive-by downloads occur when a user visits a website and the website downloads malicious code to the user's system without the user's knowledge. In some cases, the website is owned by a malicious attacker. In other cases, the attacker has exploited a vulnerability in a legitimate website and uploaded malicious code, such as through an XSS attack.

> **EXAM TIP** Two primary protections against mobile code are the use of code signing and sandboxes. Code signing is done by the website developer, and sandboxing is used on end-user systems. Code signing uses a certificate that verifies the identity of the author and also verifies that the code has not been modified since the author released it. A sandbox is an isolated area where the code can run and be tested without affecting the computer. Many web browsers now include sandbox features to isolate web-based content, but attackers are constantly searching and developing methods to bypass them.

Security issues in source code (e.g., buffer overflow, escalation of privilege, backdoor)

There are several potential issues with source code, including buffer overflows, stack overflows, escalation of privileges, and backdoors. Input validation, testing, and code reviews are important steps to prevent or detect these issues prior to releasing the application.

A buffer overflow occurs when an application receives more or different input than it expects and causes the application to fail with an unhandled error. The unhandled error exposes areas of memory that would not normally be accessible. Ideally, errors are graceful and simply give the user an error message without crashing the system. An unhandled error can crash the application or the operating system.

True or false? A stack overflow error can occur if an application tries to write more data into memory than the application has available.

Answer: *True*. Many buffer overflow errors are also called stack overflow errors. A stack overflow error occurs when an application tries to use more memory than it has available.

Applications often store temporary data onto an area of memory referred to as a stack. Each value is stacked on top of the previous value, and when the data is needed, the application pops it off the stack. When the application stores more data on the stack than the stack allows, the data overflows.

In a buffer overflow or stack overflow attack, an attacker sends a string of data to cause the buffer overflow with malicious code appended to the end. Often, this starts with a long string of no operation codes (NOOP or NOP) called a NOOP sled. The NOOP sled causes the buffer overflow, which provides access to memory that is normally restricted. The malicious code is then inserted into memory.

> **EXAM TIP** A primary method of protecting against buffer overflow and stack overflow attacks is using input validation techniques. Input validation checks the input to ensure that it is valid before it is used. For example, if a user is expected to enter a first name, the input might be limited to no more than 15 characters. A NOOP sled with malicious code would be much longer than 15 characters, considered invalid, and rejected.

Applications, system services, or daemons that have administrative or root-level privileges have an additional risk. If an attacker can cause an overflow in one of these processes, any malicious code inserted will also have elevated privileges. From a larger perspective, this is another reason to enforce the principle of least privilege even from within applications or services. If the applications or services have limited privileges, attackers will have limited privileges if they can discover and exploit a buffer overflow error.

Backdoors are often used for debugging during development. These allow developers to observe the code and variables during testing, but they should not be included in the released version.

Configuration management

Configuration management ensures that an application or system is deployed in a secure state. Combined with an aggressive change management system, it ensures that the application stays in a secure state throughout its lifecycle.

True or false? Configuration management is primarily implemented in the design phase of the system development lifecycle.

Answer: *False*. Configuration management is primarily implemented in the operations and maintenance phase of the system development lifecycle.

> **EXAM TIP** Change and configuration management ensure that a system stays in a secure state and that continuous monitoring identifies issues as they occur.

Can you answer these questions?

You can find the answers to these questions at the end of this chapter.

1. An application fails, and a user can determine details of the database management system by reading the extensive error message. What two things should be corrected?
2. What is the purpose of code signing from the perspective of the user?
3. Name two primary methods used to prevent buffer overflow attacks.
4. What is the primary purpose of configuration management?

Objective 4.3: Assess the effectiveness of software security

The goal of software development security is to ensure that software is developed in a secure manner. Assessment techniques are used to ensure that this goal has been met. An important assessment method is a code review that can uncover many different types of problems before the software is deployed.

Exam need to know...

- Assessment methods
  For example: What is included in a source code review? What is used to identify open ports on a computer?

Assessment methods

Assessment methods ensure that standard security practices have been used in the development process. They include reviews and testing prior to releasing the application and auditing after it has been released. If the system requires a formal certification, the maintenance stage can also include periodic recertification.

True or false? Because a security assessment is completed at the end of a deployment cycle, a source code review is not required.

Answer: *False*. A source code review is an important element of a security assessment and is included as part of a security assessment.

A source code review looks for a variety of security issues, including the following:

- **Input validation** This ensures that all data is checked for validity before being used. When input validation is not used, the code is highly susceptible to a wide variety of attacks, such as buffer overflow, SQL injection, and cross-site scripting attacks.

    EXAM TIP One of the ways that cross-site scripting attacks can be prevented is by checking for HTML or scripting tags. These use the less than (<) and greater than (>) characters, but these characters are not needed for typical data input.

- **Bounds and boundary checking** Bounds and boundary checking (sometimes referred to as limit checks) verifies that an application will not fail if it receives a value outside the expected boundary. It also verifies that the application operates correctly at each boundary. For example, if an application expects a value between 1 and 10, bounds checking would check values of 0, 1, 2, 9, 10, and 11. The values 1, 2, 9, and 10 are valid values and should be accepted. The values 0 and 11 are incorrect values and should be managed gracefully with error handling, providing an error message to the user without crashing the program.

- **Unauthorized Easter eggs** An Easter egg is a hidden capability of an application that can be accessed by using specific key combinations. For example, in some games you can type a code and fill your coffers with gold. These are authorized features designed to provide additional depth for gamers, and they are tested with the rest of the software. In contrast, unauthorized Easter eggs haven't gone through testing and can result in security issues.

- **Backdoors** Assessments verify that backdoors are not included in the released version. When backdoors are released, they are often discovered by attackers and exploited.

- **Malicious code** Developers can insert malicious code, such as a logic bomb, which can be discovered through a code review. A logic bomb is malicious code that runs when a condition is met, such as a specific date and time.

Assessments also include different types of testing, such as module or unit testing, logic testing or dynamic testing, integration testing, host or system testing, and network testing. A module test will validate the processes of an individual unit in a controlled environment. Integration testing verifies that modules work together.

EXAM TIP Modules are also tested for cohesion and coupling, with the goal of having high cohesion and low coupling. A module with high cohesion will perform a single task or a group of similar tasks and can easily be used by many other modules. A module with low coupling is not dependent on many other modules to complete a task.

True or false? An organization will perform fingerprinting as part of the testing for an Internet-based application.

Answer: *True*. This can be considered part of the host or system testing. Attackers use fingerprinting techniques to gain as many details about an application and its host as possible. This includes using IP and port scanners and sending follow-up messages to identify the operating system and application.

Tools such as Nessus and Nmap include port scanners to identify open ports. Open ports identify the services and protocols that are likely running on the system, based on well-known ports. For example, if port 80 is open, the server is likely a web server because port 80 is used for Hypertext Transfer Protocol (HTTP).

A fingerprinting attack follows up with specially crafted messages that are analyzed to identify the operating system. These can often identify what service packs or updates are installed.

Based on the operating system (such as Microsoft Windows or UNIX), they follow up with additional messages to identify the applications associated with the open ports. For example, Windows-based systems might run Internet Information Services (IIS) as the web server whereas UNIX systems might run Apache as the web server. By analyzing the responses, attackers can sometimes identify the version of the application and sometimes identify which application updates have been installed.

EXAM TIP Attackers use fingerprinting to identify details about Internet-based applications. Personnel performing assessment testing can use these same techniques to identify potential vulnerabilities. If a tester succeeds with fingerprinting techniques, an attacker can, too. Developers can use this knowledge to modify the code and thwart attempts by outsiders to fingerprint systems accessible on the Internet.

Can you answer these questions?

You can find the answers to these questions at the end of the chapter.

1. What is bounds checking?
2. What information can be discovered through fingerprinting?

Answers

This section contains the answers to the "Can you answer these questions?" sections in this chapter.

Objective 4.1: Understand and apply security in the software development lifecycle

1. Some software development models are the waterfall, spiral, prototyping, rapid application development (RAD), and incremental models.
2. There are two beneficiaries of CMMI ratings. First, a company can determine its level of maturity based on these levels as well as its ability to repeat its successes. More importantly, outside organizations can determine the

reliability of an organization based on its CMMI rating. Organizations with higher ratings are more likely to repeat their successes within budget and on time. An organization with a lower CMMI rating might be able to repeat a success but will often be over budget and take longer than scheduled.

3. When change management procedures are not followed, the result is loss of availability of computing resources. When unauthorized software changes are implemented, they can often impact the production schedule.

Objective 4.2: Understand the environment and security controls

1. The application should use better error handling to prevent errors from causing the application to fail. The application should be modified so that verbose messages are not sent to the user.
2. It verifies the identity of the author (or the author's company), and it verifies that the code has not been modified since it was released.
3. Buffer overflow errors can often be prevented with input validation techniques and by keeping a system up to date with current updates. Input validation ensures that the data is valid prior to sending it to the application. When buffer overflow errors are detected, the vendor releases an update to correct the error, and this update should be applied to keep systems up to date.
4. Configuration management practices ensure that systems are deployed in a consistent, secure state and remain in a secure state throughout their lifetime.

Objective 4.3: Assess the effectiveness of software security

1. Bounds and boundary checking checks routines that expect data within a certain range. It checks the values at and near the boundaries of the expected data.
2. Fingerprinting can identify open ports and likely services or protocols running on a system or host. It can also identify the operating system and the type of applications running on a system.

CHAPTER 5

Cryptography

The Cryptog Internet Protocol (IP) raphy domain covers a wide range of cryptography concepts. Three core security principles are related to confidentiality, integrity, and availability. Cryptography is closely related to the goals of preventing the loss of confidentiality and integrity. Encryption of data helps preserve confidentiality, and hashing techniques help verify the integrity of data. When preparing for this domain, you'll need to have a good understanding of basic cryptography fundamentals, along with an in-depth understanding of some underlying topics.

This chapter covers the following objectives:

- Objective: 5.1: Understand the application and use of cryptography
- Objective: 5.2: Understand the cryptographic lifecycle (e.g., cryptographic limitations, algorithm/protocol governance)
- Objective: 5.3: Understand encryption concepts
- Objective: 5.4: Understand key management processes
- Objective: 5.5: Understand digital signatures
- Objective: 5.6 :Understand non-repudiation
- Objective: 5.7: Understand methods of cryptanalytic attacks
- Objective: 5.8: Use cryptography to maintain network security
- Objective: 5.9: Use cryptography to maintain application security
- Objective: 5.10: Understand Public Key Infrastructure (PKI)
- Objective: 5.11: Understand certificate related issues
- Objective: 5.12: Understand information hiding alternatives (e.g., steganography, watermarking)

Objective 5.1: Understand the application and use of cryptography

For this exam objective, you must know the basics of how to protect the confidentiality of data. This includes data at rest (stored on a disk or other media) and data in transit (being transmitted from one system to another). The basic method applies equally to both; you use encryption to protect the confidentiality of data.

While different cryptographic methods are used in both cases, the overall purpose is simply to encrypt it to prevent the loss of confidentiality.

Exam need to know...

- Data at rest (e.g., hard drive)
  For example: How can you protect data at rest from loss of confidentiality? What are common locations where data at rest is stored?
- Data in transit (e.g., on the wire)
  For example: What are common protocols used to encrypt data transmitted over a network?

Data at rest (e.g., hard drive)

Encryption methods are used to protect the confidentiality of data, including data at rest. Data at rest is any data that is stored on a hard drive, tape, or any type of removable media, such as a USB flash drive, CD, or DVD. Data can be stored as individual files or as a group of records stored in a database.

True or false? Files can be encrypted individually on a disk using Advanced Encryption Standard (AES).

Answer: *True.* AES is a symmetric encryption algorithm used in many different cryptographic applications.

In addition to encrypting individual files, it's also possible to encrypt entire hard drives or entire volumes within a multiple-volume hard drive. Many laptop computers include a Trusted Platform Module (TPM), which can store cryptographic keys and encrypt/decrypt hard drives. Microsoft systems support BitLocker Drive Encryption, which can be used with or without a TPM to encrypt an entire volume.

> **EXAM TIP** Data at rest is any data stored on any type of media. The best way to protect against loss of confidentiality is to encrypt the data.

> **MORE INFO** Windows-based systems use New Technology File System (NTFS), which includes the Encrypting File System (EFS) to encrypt individual files. Third-party tools such as TrueCrypt (*http://www.truecrypt.org/*) also support encrypting individual files.

Data in transit (e.g., on the wire)

Data in transit refers to any data being transmitted over a medium. In general, this refers to data transferred over a wire, such as a network cable, but it also includes wireless transmissions.

Just as you can preserve confidentiality of data at rest by encrypting it, you can preserve confidentiality of data in transit by encrypting it. A variety of different encryption protocols are used to encrypt data transferred over a network.

True or false? Secure Shell (SSH) can be used to encrypt File Transfer Protocol (FTP) traffic as Secure FTP (SFTP).

Answer: *True.* SSH can be used to encrypt a wide variety of traffic, such as Telnet, Secure Copy (SCP), and SFTP. SSH is often used instead of Telnet for remotely administering Unix and Linux systems.

Some of the other protocols used to encrypt data in transit include the following:

- Secure Sockets Layer (SSL) is used to encrypt traffic transmitted over the Internet.
- Transport Layer Security (TLS) is the designated replacement for SSL, and it is used in many of the same applications as SSL.
- Internet Protocol security (IPsec) is used to encrypt traffic within a network and over public networks such as the Internet.

EXAM TIP Data in transit refers to any data being transmitted from one system to another. Encrypting data at rest does not ensure that it is encrypted when it is in transit because it is common for a system to decrypt data at rest before transmitting it. For example, a file encrypted with EFS on a server is decrypted before it is transmitted. If the data should be encrypted while in transit, protocols such as SSH, SFTP, SCP, SSL, TLS, or IPsec should be used.

MORE INFO SSL, TLS, and IPsec are discussed later in this chapter in Objective 5.8.

Can you answer these questions?

You can find the answers to these questions at the end of the chapter.

1. What is the definition of data at rest?
2. What is the primary method used to protect data at rest from the loss of confidentiality?
3. What protocol is commonly used to encrypt FTP traffic as SFTP?

Objective 5.2: Understand the cryptographic lifecycle (e.g., cryptographic limitations, algorithm/protocol governance)

Ideally, cryptographic systems will last for decades, but they rarely do. Instead, there is a balance between creating algorithms that do not require excessive processing power for legitimate users while also ensuring that they are sufficiently complex. If the algorithm requires too much processing time, it can slow down the user and impact productivity. If it is too simple, it can easily be cracked by attackers. Due to the speed of computer improvements, this is a moving target and the lifecycle of any cryptographic system is finite. The best cryptographic systems used years ago are often trivial when compared to the processing capabilities of current systems.

Exam need to know...

- Cryptographic limitations
  For example: When comparing cryptographic keys, what is the difference between keys used 10 years ago and keys used today?
- Algorithm/protocol governance
  For example: What is the function of NIST?

Cryptographic lifecycle

Computer processing power continues to increase, which regularly impacts the relative strength of cryptographic algorithms. Algorithms used ten years ago might have taken more than one hundred years of processing power to crack using computers at that time. However, the same algorithms might easily be cracked within a few hours using today's computers.

True or false? Graphics processing units (GPUs) within a system are often used in cryptographic attacks.

Answer: *True.* GPU processing power is faster than a typical central processing unit (CPU) and is often used for cryptographic attacks. As computers become faster and faster, it becomes easier to decrypt data that was encrypted using older encryption methods. With this in mind, cryptographic algorithms need to be updated regularly.

Older cryptographic algorithms used smaller cryptographic keys than newer cryptographic algorithms. It was not feasible to use very long cryptographic keys to encrypt and decrypt data with older processors because it took too long to encrypt and decrypt the data. Today's processors can easily handle these longer keys, and a key method of increasing security with cryptographic protocols is to use longer keys.

It's worth noting that keys considered to be strong today might not be very strong in just a few years.

> **EXAM TIP** GPUs are more powerful than typical CPUs, and attackers often use GPUs in cryptographic attacks. For example, password-cracking tools often make use of GPUs to decrease the time needed to crack a password.

Algorithm/protocol governance

Having a single governing body control and dictate the use of all cryptographic algorithms and protocols would provide a single standard for cryptography. However, this centralized governing body would also present many risks. For example, a single country that controlled all the cryptography standards might implement standards based on the values of the current government but not necessarily the values of other countries.

True or false? The United States National Institute of Standards and Technology (NIST) is the directive source for approved cryptographic technologies.

Answer: *False.* NIST conducts a significant amount of research and freely publishes its results, but it provides direction only to United States government agencies.

Many other organizations around the world use this research and follow NIST best practices and principles, but they are not required to do so.

NIST has defined standards for many cryptographic algorithms and published them. One of the benefits of using NIST standards for any organization is that they are freely available and created with relative transparency.

> **MORE INFO** NIST has published many research papers in its 800 series of Special Publications, which are available here: *http://csrc.nist.gov/publications/PubsSPs.html*. Additionally, NIST has published many Federal Information Processing Standards (FIPS) available here: *http://csrc.nist.gov/publications/PubsFIPS.html*. A primary benefit of these resources is that they are freely available, and NIST cryptography research is known to be open and transparent. There are additional international resources, such as ISO standards (*http://www.iso.org/*), although these are not freely available.

Can you answer these questions?

You can find the answers to these questions at the end of the chapter.

1. What is different about cryptographic keys used today than cryptographic keys used 10 years ago?
2. What are the primary types of publications published by NIST?
3. What are differences between standards published by NIST and standards published by ISO?

Objective 5.3: Understand encryption concepts

Before digging into the details of various cryptographic procedures, it's important to understand many of the basic foundational concepts related to cryptography. This section covers many of the core principles related to symmetric and asymmetric cryptography, and how hashing algorithms are used to create message digests to verify integrity. These concepts are extremely important to understand before you can fully grasp other concepts, such as how symmetric and asymmetric cryptography work together and how a digital signature is created and used.

Exam need to know...

- Foundational concepts
  For example: What is encrypted data called? What are the two elements of any cryptographic process?
- Symmetric cryptography
  For example: How many keys does symmetric cryptography use to encrypt and decrypt a single piece of data? What is the speed of symmetric cryptography when compared to asymmetric cryptography?
- Asymmetric cryptography
  For example: How many keys does asymmetric cryptography use to encrypt and decrypt a single piece of data? How is an asymmetric cryptography key distributed?

- Hybrid cryptography
  For example: What type of encryption method is used by SSL? When asymmetric and symmetric cryptography methods are used together, what is encrypted with asymmetric cryptography?
- Message digests
  For example: How large is a message digest that is created with Message Digest 5 (MD5)? What is the difference between a message digest and a number?
- Hashing
  For example: Do hashing algorithms such as MD5 and Secure Hashing Algorithm 1 (SHA1) use a cryptographic key? How does hashing provide integrity?

Foundational concepts

While cryptography has a lot of technical depth, there are some core foundation concepts that provide some basics. Mastering these basics will help you correctly answer many questions on the CISSP exam.

True or false? A primary method of ensuring confidentiality of data is to use hashing.

Answer: *False.* A primary method of ensuring confidentiality of data is to use encryption methods. Hashing is used to verify the integrity of data.

Figure 5-1 shows the basic process of encryption and decryption. Plaintext data is readable, and an encryption algorithm scrambles it in such a way that it is unreadable. The resulting text is called ciphertext data. Ciphertext data can be decrypted to create the original plaintext data.

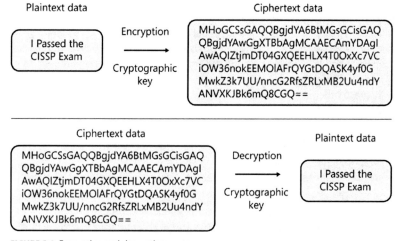

FIGURE 5-1 Encryption and decryption process.

True or false? Most encryption methods use an encryption algorithm and a key.

Answer: *True.* Both an encryption algorithm and a cryptographic key are used for most encryption methods. The encryption algorithms are published and remain constant, and a cryptographic key provides variability for the algorithm.

The following statements outline many of the generic foundational concepts related to cryptography:

- Encryption is used to preserve the confidentiality of data. Plaintext data is encrypted and becomes ciphertext data. Ciphertext data is decrypted to create the original plaintext data.
- Most encryption methods use an encryption algorithm and a cryptographic key. Longer keys used with the same encryption algorithm make it more difficult for unauthorized entities to decrypt the data.
- An encryption algorithm is constant and does not change. For example, the Advanced Encryption Standard (AES) uses a specific algorithm, and this is the same algorithm that is always used with AES. Also, most encryption algorithms are publicly available, exposing them to vigorous peer review.
- Encryption keys are not constant. For example, each time AES encrypts a file or other data it will use a different cryptographic key.
- Hashing methods are used to verify integrity.
- Hashing algorithms do not use a cryptographic key.

EXAM TIP Preventing the loss of confidentiality, integrity, and availability (CIA) are three core security goals. Encryption is directly related to preventing the loss of confidentiality. Hashing is directly related to ensuring the integrity of data.

Symmetric cryptography

Symmetric cryptography uses the same key to encrypt and decrypt a piece of data. For example, if data was encrypted with a key of 123, the same key is used to decrypt it, as shown in Figure 5-2. (In actual practice, keys will be much more complex than a simple key of 123.)

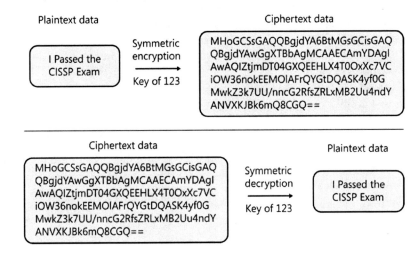

FIGURE 5-2 Symmetric encryption and decryption process.

True or false? The primary challenge with symmetric encryption is privately sharing the key.

Answer: *True*. The symmetric encryption key needs to be known by the entity encrypting the data and by the entity decrypting the data. However, it should not be known to any other entities. If any other entities discover the key, they can decrypt the data.

> **EXAM TIP** When using symmetric cryptography, the key must be transmitted privately between the two parties and changed often. If the same key is used too often, a frequency analysis attack can discover the key and access all data encrypted with the key.

Some common symmetric encryption algorithms include the following:

- **Data Encryption Standard (DES)** This is an older standard that was used in most applications for many years. However, the 56-bit key size makes it relatively easy to crack with current computers.
- **Triple DES (3DES)** This was designed as an alternative to DES using 56-bit, 112-bit, or 168-bit keys. It apples the DES algorithm three times. While it is secure, it takes more processing power than some other alternatives.
- **Advanced Encryption Standard (AES)** The U.S. government selected an algorithm formally known as Rijndael as the primary symmetric encryption standard. It was selected after NIST completed an intensive five-year evaluation process of various encryption algorithms. AES uses 128-bit, 192-bit, or 256-bit keys and takes less processing power than 3DES. AES is a popular symmetric encryption standard used to encrypt bulk data.
- **RC4 (also called arc 4)** This is named after its creator, Ron Rivest, and is sometimes called Rivest's Code. It use key sizes of 40 bits to 2,048 bits and is the symmetric encryption algorithm used with SSL.
- **International Data Encryption Algorithm (IDEA)** This was also designed as a replacement for DES. It is a block cipher that uses 128-bit keys to encrypt 64-bit blocks. It was patented but is available for free for non-commercial use.
- **Blowfish** This is a block cipher that uses variable key sizes from 32 bits to 448 bits. It creates 64-bit encrypted blocks. Blowfish runs through the encryption algorithm more than 500 times when it first creates a set of keys and subkeys. While it is a strong cipher, it takes a lot of processing power.
- **TwoFish** This is a modified version of Blowfish using keys up to 256 bits and block sizes of 128 bits. It was one of the NIST finalists in the AES competition.

True or false? AES is a stream cipher.

Answer: *False*. AES is a block cipher. It divides the data into 128-bit blocks and encrypts each block. RC4 is a stream cipher.

EXAM TIP Block ciphers encrypt fixed-size blocks of data. Cipher Block Chaining (CBC) uses data in the previous block of text to encrypt the following block. Successful decryption of any of the blocks is dependent on first decrypting all preceding blocks in the chain. In contrast, Electronic Code Book (ECB) encrypts each block of data independently.

Stream ciphers encrypt individual bits in a stream of data. An important principle that must be followed when using a stream cipher is that the seed value used to create cryptographic keys must never be used twice. This was one of the many failings of Wired Equivalent Privacy (WEP), which allowed attackers to crack it.

MORE INFO Modes of operation for block ciphers are identified in NIST SP 800-38A. A newer version is currently in draft form as SP 800-38F. Both can be accessed from the NIST PS page: *http://csrc.nist.gov/publications/PubsSPs.html*.

Asymmetric cryptography

Asymmetric cryptography uses two keys, known as a public key and a private key. There are some important but basic concepts related to these keys that you should understand.

True or false? In asymmetric cryptography, a public key is always matched with a private key.

Answer: *True*. Asymmetric keys are created as matched pairs. Data encrypted with a private key can be decrypted only with the matching public key. Similarly, data encrypted with a public key can be decrypted only with the matching private key.

A public key is freely shared with others, but a private key is always kept private. Only the owner of the key pair has access to the private key, with the possible exception of a recovery agent. Public keys are embedded within certificates and shared with others by sharing the certificate.

EXAM TIP Asymmetric cryptography is sometimes called public key cryptography or public/private key cryptography. In contrast, symmetric key cryptography is sometimes called session key cryptography, secret key cryptography, or even private key cryptography. However, calling it private key cryptography confuses it with asymmetric cryptography for many people. Asymmetric cryptography always uses a matched key pair (a public key and a private key), but symmetric key cryptography always uses a single key that is kept secret.

Figure 5-3 shows the overall process for asymmetric encryption and decryption. Keys are much more complex than 123 and 456, but for the example, assume that 123 and 456 have been created as a matched pair as a public key and a private key. A key point to remember is that data encrypted with the public key can be decrypted only with the matching private key. Similarly, if data was encrypted with the private key, it can be decrypted only with the matching public key.

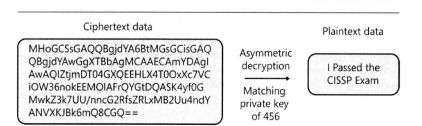

FIGURE 5-3 Asymmetric encryption and decryption process.

EXAM TIP Public and private keys are created as matched pairs. Private keys are always kept private and never shared. Public keys are shared in certificates.

True or false? Asymmetric cryptography is faster than symmetric cryptography.

Answer: *False*. Symmetric cryptography is as much as 100 times faster than asymmetric cryptography. This is one of the reasons that asymmetric cryptography is used to encrypt only the symmetric key and not entire blocks of data.

EXAM TIP Asymmetric cryptography is often used only to securely exchange a symmetric key. After both parties have the symmetric key, data is encrypted and decrypted with this symmetric key. Asymmetric cryptography takes a significant amount of processing power to encrypt and decrypt, but when it is used only to encrypt/decrypt a key, it reduces the overall processing power requirements.

RSA (named after its designers: Rivest, Shamir, and Adleman) is a popular asymmetric algorithm. The public and private keys are derived by first multiplying two large prime numbers. While it's easy to multiply two numbers, it is extremely complex to factor the product of these two large prime numbers.

EXAM TIP Mathematically, it is difficult to factor the product of two large prime numbers. RSA takes advantage of this by starting with two large prime numbers to create the public and private keys. When the keys are sufficiently large, it is not feasible to detect the original prime numbers in a reasonable amount of time.

MORE INFO RSA laboratories sponsored the RSA Factoring Challenge several years ago, which encouraged cryptographers to identify the factors of large prime numbers. Many large numbers (up to 768 bits) have been factored, although they often take hundreds of computing years to complete. It's estimated that the RSA-2048 factor (using 2,048 bits) will likely not be factored for many more decades. You can read about the RSA Factoring Challenge here: *http://www.rsa.com/rsalabs/node.asp?id=2094*.

True or false? Elliptic curve cryptography (ECC) is less efficient than typical asymmetric encryption methods.

Answer: *False*. ECC takes less processing power because it is more efficient than typical asymmetric encryption methods.

EXAM TIP ECC is commonly used in smaller mobile devices because it requires less processing power.

Diffie-Hellman and El Gamal are two additional asymmetric cryptography methods. These methods use discrete logarithms and can be used to privately share a symmetric key over a public network.

Hybrid cryptography

Hybrid cryptography refers to using a combination of symmetric and asymmetric cryptography. The most common way this is done is by encrypting a symmetric key by using asymmetric cryptography. This allows two entities to privately share a symmetric key without any previous coordination.

EXAM TIP When symmetric and asymmetric cryptography are combined, the primary purpose of asymmetric cryptography is to privately share the symmetric key. The symmetric key is used to encrypt the data because it is much more efficient than asymmetric cryptography.

MORE INFO The section on SSL and TLS in Objective 5.8 later in this chapter shows how symmetric and asymmetric cryptography works together in Hypertext Transfer Protocol Secure (HTTPS) sessions.

Message digests

A message digest is a fixed string of characters that is created after applying a hashing algorithm to a message or a file.

True or false? A message digest is simply a number derived from a message or a file.

Answer: *True*. A message digest is also known as a hash, and the hash is a number. Hashes are typically represented by using hexadecimal numbers, which include the letters A through F.

Two popular message digest algorithms are Message Digest 5 (MD5) and Secure Hashing Algorithm 1 (SHA1). Figure 5-4 shows the process for creating a hash. Notice that a key is not used with a hashing algorithm. Instead, the hashing algorithm is applied to either the message or the file to create the message digest.

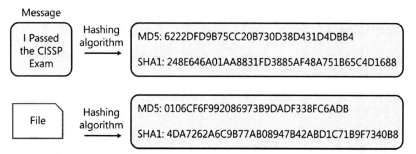

FIGURE 5-4 Creating a message digest with hashing algorithm.

EXAM TIP MD5 creates 128-bit hashes. When expressed in 4-bit hexadecimal, it creates a 32-character hexadecimal string. SHA1 creates 160-bit hashes, commonly expressed as 40-character hexadecimal strings.

NOTE A *checksum* is another name for a message digest or a hash, and it is often used to verify the integrity of packets or frames sent over a network. A checksum is calculated against a message and is included with the message. When it is received, the checksum is recalculated and compared against the received checksum.

MORE INFO MD5 is documented in RFC 1321 and RFC 6151, available here: *http://tools.ietf.org/html/rfc1321* and *http://tools.ietf.org/html/rfc6151*. NIST has published FIPS 180-2 "Secure Hash Standard," which describes the different iterations of SHA algorithms. You can access it here: *http://csrc.nist.gov/publications/fips/fips180-2/fips180-2.pdf*.

Hashing

Hashing is the process of creating a fixed string of characters (called a message digest or the hash) from a message or file. MD5 and SHA1 are two common hashing algorithms. SHA2 creates message digests of 224, 256, 384, or 512 bits. When 256 bits are used, it is commonly called SHA-256.

True or false? The primary purpose of a hash is to prevent loss of confidentiality.

Answer: *False*. The primary purpose of hashing is to verify integrity. Encryption prevents loss of confidentiality.

A hashing algorithm will always create the same hash when executed against a piece of data. If the data is changed, the hashing algorithm will create a different hash. Imagine that you run the algorithm against a file and get a hash of 1A2B3C on Monday. If you use the same algorithm against the same file on Tuesday and get a hash of 1A2B3C, you know that the original file has not changed because the hashes are identical. Imagine that you run the same algorithm against the file on Wednesday but get a different hash, such as 3D2E1B. With this information, you know that the file is different because the hashes are different.

Figure 5-5 shows one way this is done to provide integrity. Imagine that you want to download a trial copy of one of the Microsoft Server products. If you use TechNet, you'll see a link to download the file and you can also view the original SHA1 hash. After you download the file, you can calculate the hash on the downloaded file. If the file has not been modified, both hashes should be the same. In the figure, you can see that the two hashes are different, indicating that the file has lost integrity.

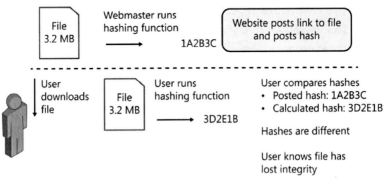

FIGURE 5-5 Hashing is used for integrity.

There are many reasons why the file might be different. It might have been infected with malware, or it's possible that it just lost some bits during the download. The different hashes don't tell you why the file is different, but when you know it has lost integrity, you know you shouldn't use it.

> **EXAM TIP** Hash functions such as MD5, SHA-1, and SHA-256 create a fixed string of characters and are used to verify integrity. As long as the original data remains the same, the resulting hash is always the same. Hashing is also called one-way encryption or a one-way function. You can create the hash from the original data, but you cannot determine the original data from just the hash. In contrast, ciphertext data can be decrypted to re-create the original plaintext data but you cannot use the hash to re-create the original data because a hashing function is one-way.

> **MORE INFO** There are several free utilities available to calculate hashes. Microsoft has published the File Checksum Integrity Verifier (FCIV), which you can use from the command line to calculate MD5 and SHA1 hashes. You can get it by searching "File Checksum Integrity Verifier" on the Microsoft download site (*http://www.microsoft.com/download*).

True or false? A hashed message authentication code (HMAC) is used for authentication of the sender.

Answer: *False*. An HMAC is used for integrity. In this context, authentication refers to the authenticity of the message, not authentication of the sender.

HMAC adds security to a hashing algorithm by encrypting the resultant hash with a symmetric key. Common HMAC algorithms are HMAC-MD5 and HMAC-SHA1, and they provide better security than MD5 or SHA1 alone.

> **EXAM TIP** An HMAC provides data integrity, but it does not provide authentication of the sender or non-repudiation because it uses symmetric cryptography. In contrast, a digital signature (covered later in this chapter) provides integrity, authentication of the sender, and non-repudiation—the sender cannot deny sending the message.

Can you answer these questions?

You can find the answers to these questions at the end of the chapter.

1. What changes to the cryptographic key results in the overall encryption becoming more complex?
2. How many keys are used with symmetric cryptography for any given piece of data?
3. What key is never shared in asymmetric cryptography?
4. What key is used to decrypt the session key in an HTTPS session?
5. How many bits are in a SHA1 message digest?
6. If a hashing algorithm creates two different hashes from what is supposed to be the same data, what do you know?

Objective 5.4: Understand key management processes

Key management refers to the different methods to create, distribute, store, destroy, and recover keys during their lifecycles. Many automated methods are available for key management, and when evaluating them, it's important to understand the basic security requirements.

Exam need to know...

- Creation/distribution
  For example: What is a key derivation function used for? What type of encryption does Kerberos use when encrypting keys?
- Storage/destruction
  For example: What level of protection is required for cryptographic keys?
- Recovery
  For example: When referring to recovery in the key management process, what two things can be recovered?
- Key escrow
  For example: What level of security should be used to protect keys placed in escrow?

Creation/distribution

Cryptographic keys need to be sufficiently random and sufficiently long to prevent an attacker from easily identifying them. When symmetric keys are used, it's common to use a different key for each piece of data that is being encrypted and decrypted. For example, File A is encrypted and decrypted with Key A, and File B is encrypted and decrypted with Key B. However, Key A cannot be used to decrypt File B. Similarly, Key B cannot be used decrypt File A.

True or false? A key derivation function (KDF) is used to create matched public and private keys.

Answer: *False*. A KDF is a master key used to create symmetric keys. For example, a database server might have a master key and then use a KDF to create different symmetric keys for entire databases, tables, individual columns, and/or other database objects.

> **EXAM TIP** KDFs ensure that keys are different and sufficiently random to prevent successful cryptanalysis attacks. They typically use a master key, such as a password, and combine it with a randomness factor, such as a random number.

When creating matched keys for asymmetric cryptography, applications are often used to create them automatically.

True or false? Cryptographic keys are created with algorithms and pseudorandom number generators.

Answer: *True*. Deterministic (or constant) algorithms are combined with a random seed value to create pseudorandom keys. They start with a random seed value to add an element of randomness.

> **MORE INFO** NIST has published a draft copy of SP 800-133 titled "Recommendation for Cryptographic Key Generation." This document includes a list of definitions, acronyms, and symbols that are valuable for understanding many cryptographic topics. You can access it from this page: *http://csrc.nist.gov/publications/PubsSPs.html*.

Kerberos is a network authentication protocol that includes a Key Distribution Center (KDC) to distribute keys. It was originally developed at MIT and is widely used. Microsoft's Active Directory uses Kerberos.

True or false? Kerberos 5 uses AES to securely transmit keys.

Answer: *True*. Kerberos version 5 includes a KDC, and keys are encrypted with AES to protect them.

> **MORE INFO** Microsoft has used Kerberos in several iterations of its server products, including Windows Server 2012. The following page provides an overview of Kerberos used with Active Directory, including several practical applications of Kerberos: *http://technet.microsoft.com/library/hh831553.aspx*.

Storage/destruction

An important consideration when storing cryptographic keys is that private keys must be kept private. Someone gaining access to a private key can then decrypt any data that has been encrypted with the matching public key. Similarly, if a symmetric key will be used for an extended period of time, it must also be protected.

> **EXAM TIP** The level of security for cryptographic keys should be at least as strong as the level of security for the data it is encrypting. For example, if cryptographic keys are used to encrypt top secret data, these cryptographic keys should be protected as if they are top secret data.

True or false? Cryptographic keys should be erased when they are no longer needed.

Answer: *False*. Erasing a key rarely removes it completely from a system. When a cryptographic key is no longer needed, it should be destroyed in such a way that all remnants of the key are removed.

True or false? The lifetime of a cryptographic key should be long for keys that are frequently used.

Answer: *False*. If a key is frequently used, it should have a short lifetime.

The key lifetime is also related to the value of data. Keys used to encrypt highly sensitive data should be changed more often than keys used to encrypt less sensitive data. Also, the keys used to encrypt highly sensitive data should be longer than other keys, adding to the complexity of the ciphertext.

Recovery

Part of key management is key recovery. It ensures that there is some method available for a key to be recovered if the original key is lost.

True or false? A designated key recovery agent has access to private keys.

Answer: *True*. A designated key recovery agent can either recover the original private key used to encrypt data or use an alternate private key to recover the data.

> **MORE INFO** NTFS includes EFS to encrypt files and has built-in support for a designated recovery agent. When used, EFS files can be opened with the original user's private key or the designated recovery agent's private key. Knowledge base article 241201 discusses this process: *http://support.microsoft.com/kb/241201*.

Key escrow

Key escrow refers to putting cryptographic keys in a safe place so that they can be used if the original key is no longer available.

True or false? Backing up keys and storing them in a secure location is one method of key escrow.

Answer: *True*. There are many different methods of key escrow, including making backup copies of a key. If the original key is lost or no longer accessible, the backed-up key can be retrieved and restored.

True or false? Only public keys are placed in escrow.

Answer: *False*. Public keys are publicly available and do not need to be placed in escrow. However, the private key is placed in escrow when using public and private keys.

> **EXAM TIP** Key escrow isn't a requirement but is used if data loss is unacceptable. In some cases, an organization decides not to use a key escrow to eliminate the risk of compromised keys held in escrow. They are willing to accept the risk of data loss if the original key is lost as a tradeoff against the risk of escrow compromise. When key escrow methods are used, it's important to use strong security methods to protect the location where the keys are stored.

Can you answer these questions?

You can find the answers to these questions at the end of the chapter.

1. What is the relationship between the length of cryptographic keys and the value it protects?
2. What is the relationship between the lifetime of a cryptographic key and the value it protects?
3. What are two items that can be recovered if a key recovery process is in place?
4. What types of keys are held in escrow?

Objective 5.5: Understand digital signatures

Digital signatures provide a digital alternative to an actual handwritten signature. They provide authentication of the sender, verify that the message has not been modified, and prevent senders from later denying they sent a message.

Exam need to know...

- Purpose and process of digital signatures
  For example: What are the three benefits of a digital message? How is a digital signature created?

Purpose and process of digital signatures

A digital signature is sent with an email to provide the following benefits:

- **Integrity**   Hashing techniques verify integrity of the message.
- **Authentication**   When the digital signature can be decrypted with the sender's public key, it provides authentication of the sender. If the digital signature can be decrypted with the sender's public key, it verifies that it was encrypted with the sender's private key, which is available only to the sender.
- **Non-repudiation**   Because the sender's public key can decrypt the message, it must have been encrypted with the sender's private key. The sender can't later deny sending the message.

True or false? A digital signature can be used to provide accountability.

Answer: *True*. Because the digital signature authenticates the sender and provides non-repudiation, the sender can be held accountable for the contents of an email with a digital signature.

> **EXAM TIP** A digital signature provides integrity, authentication of the sender, and non-repudiation. Digital signatures do not encrypt the actual message, but if the message needs to remain confidential, the message could be encrypted and include a digital signature. Also, digital signatures do not include a timestamp, but timestamps could be added to prevent someone from replaying the message.

True or false? A digital signature is decrypted with the sender's public key.

Answer: *True*. A digital signature is encrypted with the sender's private key and can be decrypted only with the sender's public key.

A digital signature provides several benefits, including the following:

- It provides integrity with the use of hashing.
- It provides authentication of the sender so that the recipient has assurances that it was sent from the sender's account.
- It provides non-repudiation, preventing the user from denying sending the message.

The following steps are used when creating a digital signature:

1. The sender creates an email message.
2. The sender creates a hash of the email message.
3. The sender encrypts the hash with the sender's private key.
4. The sender sends the original message and the encrypted hash of the message to the recipient. The encrypted hash of the message is the digital signature.

> **EXAM TIP** It's important to realize that encryption of a message and the use of a digital signature are two separate steps. If the message needs to remain confidential, it can also be encrypted in a separate step. However, it's also possible to digitally sign a message without encrypting the message.

5. The recipient receives the email and the digital signature.
6. The recipient retrieves the sender's public key by retrieving the sender's certificate.
7. The recipient decrypts the digital signature by using the sender's public key. This provides the hash of the original message.
   a. If the decryption succeeds, it proves the digital signature was encrypted with the sender's private key.
   b. If the decryption fails, it indicates that the digital signature has been modified or encrypted by someone else's private key.
8. The recipient calculates the hash on the received message.

9. The recipient compares the two hashes.
   a. If they are the same, the recipient knows that the received message has not been modified.
   b. If the hashes are different, the recipient knows that the received message is not the same as the sent message. The message has lost integrity.

EXAM TIP A digital signature is decrypted with the sender's public key. A digital signature is created by encrypting the hash of a message with the sender's private key. Digital signatures provide integrity, authentication, and non-repudiation.

NOTE Many security professionals mention a flaw in a digital signature's ability to provide definitive authentication and undeniable non-repudiation. Specifically, if you can decrypt the digital signature, you know that it was encrypted with the sender's private key. However, this doesn't provide incontrovertible proof that the owner of the private key actually sent the message. There is a small possibility that the owner's private key was stolen and is being used by the thief. If the owner denies sending the message, other forensic methods can be used to determine whether the sender's key has been compromised.

MORE INFO You can view some short videos related to security on YouTube (http://www.youtube.com). If you search on "Darril Gibson Security+", you'll find four videos related to digital signatures, hashing, encryption, and HTTPS and SSL. Even though these videos are related to Security+, the same concepts apply to the CISSP exam.

True or false? You can use a hashed message authentication code (HMAC) instead of a digital signature and achieve the same benefits.

Answer: *False*. An HMAC does not provide authentication or non-repudiation. An HMAC does provide integrity with the use of an encrypted hash, but it doesn't use public and private keys, so it does not provide authentication of the sender or the non-repudiation.

Can you answer these questions?

You can find the answers to these questions at the end of the chapter.

1. What are the cryptographic benefits of a digital signature?
2. What is needed to verify a digitally signed message?
3. What type of key is needed to create a digital signature?
4. Is the confidentiality of a message protected with a digital signature?

Objective 5.6: Understand non-repudiation

Non-repudiation refers to methods used to prevent someone from denying they took an action. It is an important benefit of a digital signature and is supported through public key cryptography.

Exam need to know...

- Non-repudiation
  For example: What primary method of cryptography provides non-repudiation?

Non-repudiation

One of the goals of cryptography is to prevent users from denying they took an action. Within cryptography, a digital signature provides non-repudiation.

True or false? Non-repudiation is provided when you can open a digital signature with the sender's public key.

Answer: *True*. A digital signature is encrypted with the sender's private key and can be decrypted only with the sender's public key. Public and private keys come as matched pairs. A public key from one matched pair will not decrypt anything encrypted with a private key from a different matched pair.

Another way that non-repudiation is used is with Internet documents and transactions. An individual can be granted access to an Internet-based website by using specific credentials. After the individual logs on to the site, the individual is required to read and acknowledge documents, often by checking a box. This can be used later as proof that the individual acknowledged the document.

> **EXAM TIP** A primary method of providing non-repudiation is with a digital signature. This requires that the users keep their private key private. If the private key is compromised, someone else can use it with a digital signature.

> **MORE INFO** Similarly, audit logs are used to document user activity, such as when a user logs on to a system or accesses a file. Strong authentication procedures combined with auditing also provide non-repudiation. Authentication and auditing were covered in Chapter 1, "Access Control."

Can you answer these questions?

You can find the answers to these questions at the end of the chapter.

1. How does a digital signature provide non-repudiation?
2. What is required for an audit trail to provide non-repudiation?

Objective 5.7: Understand methods of cryptanalytic attacks

Ideally, only authorized users can decrypt and read encrypted data. However, attackers use a variety of different cryptanalytic attack methods to decrypt data. When they are successful, the confidentiality of the data is lost. This section covers many of the common methods used by attackers, although it's worthwhile mentioning that cryptanalysis is a moving target. As cryptographic methods become stronger, cryptanalytic methods used to crack them also become better.

Exam need to know...

- **Chosen plaintext**
  For example: What does a user need to know for a chosen plaintext attack?
- **Social engineering for key discovery**
  For example: What is used to prevent someone from reading a password on a screen?
- **Brute force (e.g., rainbow tables, specialized/scalable architecture)**
  For example: What is a basic protection against rainbow tables? What is the relationship between hash sizes and collisions?
- **Ciphertext only**
  For example: What is a primary protection against ciphertext-only attacks?
- **Known plaintext**
  For example: What is the difference between a known plaintext attack and a chosen plaintext attack?
- **Frequency analysis**
  For example: What does a frequency analysis attack look for?
- **Chosen ciphertext**
  For example: What is the goal of an attacker when attempting to decrypt a small portion of ciphertext?
- **Implementation attacks**
  For example: What is a side channel attack? What environmental factors are typically analyzed in an implementation attack?

Chosen plaintext

In a chosen plaintext attack, an attacker has ciphertext data and some knowledge about the original plaintext data.

True or false? In a chosen plaintext attack, an attacker can predict likely character strings that are embedded in the original message of encrypted ciphertext.

Answer: *True.* For example, based on previous messages, the attacker might know that some character strings, such as "Subject:" and "From:", are always included in a message. The attacker uses this information to locate and decrypt those character strings and, if successful, can now decrypt the rest of the message.

In World War II, England described areas where mines were placed by using gardening terms such as fruits and vegetables. The Germans then used these fruit and vegetable words in their encrypted messages, allowing England to perform chosen plaintext attacks. Today, encouraging an entity to encrypt specific plaintext words or phrases is referred to as gardening. When these words are encrypted, a chosen plaintext attack might be successful, allowing the entire message to be decrypted.

EXAM TIP A successful chosen plaintext attacker can gain insight into the key used to encrypt the original plaintext data. If the key is discovered, it allows the attacker to decrypt the entire ciphertext message.

Social engineering for key discovery

Social engineering methods are commonly used by attackers to trick people into giving up secrets, and some common methods were introduced in Chapter 1. They often trick unsuspecting users simply by talking to them or asking questions.

True or false? Shoulder surfing is the practice of gaining information just through observation.

Answer: *True*. Shoulder surfing can be used to gain information from a user's display monitor. Using display filters to limit the viewing angle is helpful.

Encrypting data doesn't prevent a shoulder surfing attack because the data will be presented to the user in an unencrypted format.

> **EXAM TIP** A core principle related to key management states that keys should never be displayed as clear text. Sometimes passwords are used for keys, but these should be masked with a character such as the asterisk (*).

Brute force (e.g., rainbow tables, specialized/scalable architecture)

Brute force attacks attempt to try all the possibilities one by one. The attack isn't very elegant but, combined with some other methods, can be successful.

In a brute force attack, there are three steps:

1. Guess a password or string combination.
2. Calculate the hash of the guessed password.
3. Compare the calculated hash with the hash of an actual password.
   a. If it matches, the guessed password might be the actual password or might be a different character string that is creating the same hash. Either way, the guessed password can be used as the actual password.
   b. If it doesn't match, steps 1 through 3 are repeated.

Guessing a password and calculating the hash takes the most amount of processing time. A rainbow table is a table of string combinations that have been previously hashed, and when it is used, the attacker needs only to compare the password hash with each previously hashed password in the rainbow table. When a match is found, the attacker can use the lookup table to identify the original password or a string of characters that can be used instead of the original password.

True or false? Salting a hash reduces the success of rainbow tables.

Answer: *True*. A salt is a group of random bits that is added to a hash, and it is commonly used to randomize password hashes. When the password hashes are salted, rainbow tables must be significantly larger than normal, making them impractical.

> **EXAM TIP** Passwords should never be stored in clear text. Ideally, they should be stored using a salted hash to prevent an attacker from discovering them.

True or false? A birthday attack is a form of attack that identifies collisions related to passwords.

Answer: *True*. A birthday attack is a form of attack using probability theory, and the same principle is used to find collisions. It's based on the probability that in a room of 23 random people, there's a 50 percent chance that at least two people have the same birthday.

> **MORE INFO** The birthday paradox is often quoted. If you love statistics, it's probably clear to you, but if you're not a statistics fan, it might be a little fuzzy. This page explains it in simpler terms: *http://betterexplained.com/articles/understanding-the-birthday-paradox/*.

A collision is an instance when two different pieces of data have the same hash. Collisions are not desirable and represent a vulnerability for hashing algorithms. For example, passwords are sometimes stored as hashes of the password instead of the actual password. When a user enters a password, the hash of the password is compared to the stored hash. In a collision attack, the attacker attempts to identify any character string that creates the same hash. If successful, the attacker can use the character string without knowing the actual password.

> **EXAM TIP** Longer hashes decrease the probability of collisions. For example, MD5 uses 128-bit hashes and is much more susceptible to collisions than SHA-256, which creates 256-bit hashes. In this context, MD5 has a low resistance to collisions and SHA-256 has a high collision resistance.

Ciphertext only

In a ciphertext-only attack, the attacker has access only to encrypted ciphertext data. Early encryption methods could often be decrypted by using only ciphertext, combined with a pencil and paper. However, as encryption methods have become more complex, using progressively longer keys, this has become more and more difficult.

True or false? When a key is used more than once, it increases the chances that a ciphertext-only attack will succeed.

Answer: *True*. When an encryption key is used more than once, it makes it easier for statistical analysis methods to detect patterns. A successful ciphertext-only attack can decrypt the data and often discover the key.

> **EXAM TIP** Wired Equivalent Privacy (WEP) is susceptible to ciphertext-only attacks. One of the key reasons is that it uses the same keys more than once. WEP has been replaced with Wi-Fi Protected Access (WPA) and WPA2, which are used in place of WEP for wireless security.

Known plaintext

In a known plaintext attack, the attacker has a sample of plaintext data and ciphertext data. This is similar to a chosen plaintext attack except that more plaintext data is available and used in the analysis.

EXAM TIP In a known plaintext attack, the attacker has large samples of both plaintext data and the encrypted ciphertext data. In a chosen plaintext attack, the attacker focuses on specific character strings instead of the entire message.

Frequency analysis

In a frequency analysis attack, the attacker analyzes large portions of ciphertext.

True or false? Frequency attacks look for repetitions in ciphertext.

Answer: *True*. Frequency attacks look for repeated strings in ciphertext and attempt to predict what strings would be repeated to break the code.

EXAM TIP Frequency analysis attacks use different statistical analysis methods to identify patterns. One method is to compare the ciphertext with known patterns in the original language of the plaintext data. For example, the English language uses the letters R, S, T, L, N, and E more often other letters. A common method used to thwart frequency analysis attacks is to change the key often and not reuse keys.

Chosen ciphertext

In a chosen ciphertext attack, the attacker focuses on a small element of the ciphertext and attempts to decrypt it. If successful, the attacker can discover the key and then use this key to decrypt the rest of the ciphertext.

True or false? El Gamal is susceptible to chosen-ciphertext attacks

Answer: *True*. A weakness of El Gamal is that it is not secure against a chosen ciphertext attack. El Gamal is an asymmetric encryption protocol based on Diffie-Hellman. Modifications to El Gamal add padding to protect against this weakness.

Implementation attacks

Implementation attacks focus on the physical implementation of the cryptographic system. These are commonly referred to as side-channel attacks.

True or false? Attacks on smart cards include differential cryptanalysis attacks.

Answer: *True*. Differential power analysis attacks are used against smart cards to gain information about the cryptographic keys.

Attacks on smart cards are known as side-channel attacks. They are based on the premise that a cryptographic system has more attack vectors than just the algorithm and the key. Side-channel attacks typically analyze one of the following environmental factors:

- **Power** This type of attack analyzes minute differences in power consumption as a system is encrypting or decrypting data.
- **Timing** This analysis measures the amount of time it takes for the system to encrypt or decrypt data.
- **Clock** When systems have a clock as an input, side-channel attacks manipulate the clock in an attempt to cause the cryptographic system to fail.

In some cases, the output from a failure provides insight into the cryptographic key.

EXAM TIP There are a wide variety of side-channel attacks against smart cards, including differential power analysis attacks and timing analysis attacks. These side-channel attacks analyze different factors and can often gain information about the cryptographic keys.

MORE INFO Discretix Technologies, Ltd., has published white papers on side-channel attacks. You can access "Known Attacks Against Smart Cards" at *http://www.hbarel .com/publications/Known_Attacks_Against_Smartcards.pdf* and "Introduction to Side Channel Attacks" at *http://www.hbarel.com/publications/Introduction_To_Side _Channel_Attacks.pdf*.

Can you answer these questions?

You can find the answers to these questions at the end of the chapter.

1. What is a salt and how does it relate to rainbow tables?
2. What helps prevent the possibility that a ciphertext-only attack will succeed?
3. What are some common environmental factors analyzed in an implementation attack?

Objective 5.8: Use cryptography to maintain network security

A key goal of cryptography within network security is to protect data in transit. This is done by encrypting data prior to sending it over a network. In addition to understanding the protocols used to encrypt data in transit, it's also important to have a clear understanding of the differences between link encryption and end-to-end encryption.

Exam need to know...

- Understand differences between link and end-to-end encryption
  For example: What type of encryption prevents an attacker from performing traffic analysis?
- Understand protocols used for network security
  For example: What is the designated replacement for SSL? How is a symmetric key established in an HTTPS session?

Link vs. end-to-end encryption

Data sent over a network can be encrypted by using either link encryption or end-to-end encryption. The primary difference is that end-to-end encryption encrypts the payload, while link encryption encrypts both the payload and the routing data.

True or false? Link encryption is the best method to prevent an attacker from performing traffic analysis.

Answer: *True*. Link encryption encrypts the traffic on a hop-by-hop basis. If attackers intercept the data, they will be able to view the routing information between two hops (two routers) but they won't be able to view information on the original source or the ultimate destination.

> **EXAM TIP** Link encryption is often used on public networks provided by Internet service providers (ISPs). It provides confidentiality of the data and confidentiality of the data path. End-to-end encryption is typically used within a private network. It provides confidentiality of the data but does not encrypt the data path. Traffic that is transmitted through both public and private networks might use both link encryption and end-to-end encryption.

SSL and TLS

Secure Sockets Layer (SSL) has been the primary protocol used with HTTPS for decades. It was created before there were standards. Transport Layer Security (TLS) was created as an RFC standard and is the designated replacement for SSL.

True or false? SSL is a proprietary protocol.

Answer: *True*. SSL was created by Netscape. While it has been a strong protocol that has worked for many years, there wasn't a way to upgrade it because it wasn't an open standard. In contrast, TLS is an open standard and has been updated.

SSL version 3.0 is the version that has been in use since it was released by Netscape in 1996. TLS 1.0 was based on SSL 3.0 and includes the ability to automatically downgrade to SSL 3.0 if necessary. Newer versions of TLS are TLS 1.1 and TLS 1.2.

> **EXAM TIP** TLS is the designated upgrade for SSL and can be used in place of SSL in applications. It is more extensible than TLS and can be upgraded by the Internet Engineering Task Force (IETF) as a newer RFC.

> **MORE INFO** TLS 1.0 was originally defined in RFC 2246 and was most recently updated as TLS 1.2 in RFC 5246. You can access these RFCs here: *http://tools.ietf.org /html/rfc2246* and *http://tools.ietf.org/html/rfc5246*. SSL 3.0 is published as a historical document in RFC 6101 (*http://tools.ietf.org/html/rfc6101*).

True or false? SSL uses asymmetric cryptography to privately share the session key and uses symmetric cryptography to encrypt an HTTPS session.

Answer: *True*. An HTTPS session uses SSL or TLS with a combination of both symmetric and asymmetric cryptography.

Figure 5-6 shows how an HTTPS session is created.

1. The client sends the HTTPS request to begin an HTTPS session.
2. The server responds by sending the certificate, which includes the server's public key.

3. The client creates a symmetric key that will later be used for the HTTPS session. For this example, imagine that the symmetric key is a simple 3-digit number, 479, but that the actual key is more complex. At this point, only the client knows the symmetric key.
4. The client encrypts the symmetric key of 479 with the server's public key. The result for this example is IP@$$ed.
5. The client sends the encrypted symmetric key to the server. Because this was encrypted with the server's public key, only the server's matching private key can decrypt it. Further, only the server has the server's matching private key because the private key is always kept private.
6. The server receives the encrypted symmetric key and decrypts it with its private key. At this point, both the client and the server know the symmetric key (479) but no one else knows it.
7. The remaining HTTPS session is encrypted by using the symmetric key.

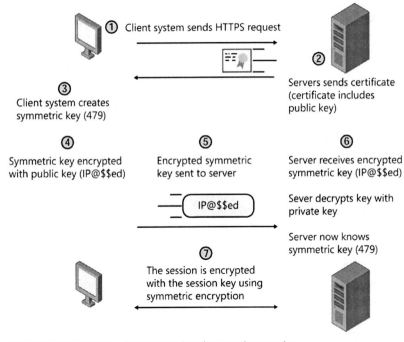

FIGURE 5-6 HTTPS session using asymmetric and symmetric encryption.

EXAM TIP SSL and TLS use a combination of both asymmetric and symmetric encryption in HTTPS sessions. Symmetric keys are not reused after a session is completed. Instead, the asymmetric/symmetric process is repeated for each HTTPS session. Even if someone was able to somehow identify the symmetric key used in one session, they cannot use this key to decrypt data from another session.

IPsec

Internet Protocol security (IPsec) can be used on internal networks and on the Internet to protect data. It uses Authentication Header (AH) for authentication of the data origin and integrity and Encapsulating Security Payload (ESP) to encrypt the payload for confidentiality.

IPsec uses separate security associations (SAs) for each secure channel. When the connection is first established, the two entities negotiate the strongest security methods that both entities can support. For example, both entities might support AES for encryption and SHA1 for integrity. The SA also includes cryptographic keys that both will use for this channel.

True or false? A single IPsec security association is used by two systems to transmit and receive data.

Answer: *False*. Two SAs will be used if two systems are transmitting and receiving. A separate SA is used for each direction.

> **EXAM TIP** IPsec is used in both public and private networks to secure traffic. It can be used with ESP to encrypt traffic while also providing data integrity. It can also be used with AH alone for integrity without encrypting the traffic.

> **MORE INFO** Chapter 2, "Telecommunications and Network Security," also discussed IPsec in the context of data communications.

> **EXAM TIP** Keys should never be transmitted over the network in clear text. A common procedure is to create a symmetric key that will be used by two entities and then use asymmetric encryption to encrypt it prior to transmitting it over the network.

Can you answer these questions?

You can find the answers to these questions at the end of the chapter.

1. What type of network encryption will encrypt the routing data?
2. What entity creates the symmetric key within an HTTPS session?
3. What is used to ensure that an HTTPS session key cannot be discovered by unauthorized entities?
4. What cryptographic key is used to encrypt data within an HTTPS session?
5. What is included in an IPsec security association?

Objective 5.9: Use cryptography to maintain application security

Applications commonly use cryptography to provide additional security. This includes using encryption to provide confidentiality and using hashing methods to provide integrity. HTTPS is one example that was explored in-depth in the previous section. Another common way cryptography is used within applications is with email.

Exam need to know...

- Using cryptography with applications
  For example: What is the primary standard used to secure email?

Application security

Secure Multipurpose Internet Mail Extensions (S/MIME) is the primary standard used to secure email sent over the Internet and internally within private networks.

True or false? S/MIME is dependent on a Public Key Infrastructure (PKI), but Pretty Good Privacy (PGP) is designed to use a web of trust.

Answer: *True.* S/MIME is dependent on the use of certificates for exchanging keys and the trust relationships associated with certificates. Different versions of PGP support different methods of exchanging keys without the use of a PKI (although some versions of PGP can use a PKI).

The original PGP code was developed by Phil Zimmermann and was designed for private individuals to communicate privately with each other. It has been bought and sold several times and is currently owned by Symantec. PGP can be used to encrypt email and use digital signatures just as S/MIME can.

> **EXAM TIP** A primary difference between S/MIME and older PGP methods is the way the keys are exchanged. Both support encryption of the email contents for confidentiality and the use of digital certificates for authentication, integrity, and non-repudiation. However, S/MIME is dependent on a PKI to distribute certificates. PGP uses other methods to distribute certificates.

> **MORE INFO** S/MIME is defined as a public standard in RFC 3851 (*http://tools.ietf.org/html/rfc5751*). Symantec (*http://www.symantec.com/pgp*) owns the rights to PGP, which it purchased from PGP Corporation, and has several commercial products supporting different PGP implementations. There are also open standard versions of PGP such as the standards managed by the The OpenPGP Alliance (*http://www.openpgp.org/*) and defined in RFC 4880 (*http://www.ietf.org/rfc/rfc4880*).

True or false? Primary reasons to use cryptography with email are to ensure confidentiality and provide authentication of recipients.

Answer: *False.* A sender cannot authenticate recipients. Primary reasons to use cryptography with email include encrypting the message for confidentiality, providing authentication of the sender, non-repudiation, and integrity.

> **EXAM TIP** Email cryptography applications focus on two methods. Encryption of the email provides confidentiality and helps prevent unauthorized disclosure. Digital signatures validate the integrity of the message, provide authentication of the sender, and non-repudiation.

True or false? Encrypted email can be decrypted by using the sender's public key.

Answer: *False*. Encrypted email can be decrypted only with the recipient's private key. If the sender's public key could decrypt it, confidentiality is lost because the sender's public key is publicly available. Anyone could decrypt it.

The following steps are used when encrypting and sending email:

1. The sender retrieves the recipient's certificate to get the recipient's public key.
2. The sender encrypts the email with the sender's public key.
3. The encrypted email is sent to the recipient. Only the recipient's private key can decrypt it. If it is intercepted, it cannot be read because only the recipient has the recipient's private key.
4. The recipient receives the email and decrypts it with the recipient's private key.

While the recipient's public key is used to encrypt and the recipient's private key is used to decrypt, this can be done by using one of the following two methods:

- **Encrypting a symmetric key with the public key.** In most cases, the recipient's public key is used to encrypt a symmetric key and the symmetric key is used to encrypt the email.
    a. The sender creates a symmetric key.
    b. The sender encrypts the email with the symmetric key.
    c. The sender encrypts the symmetric key with the recipient's public key.
    d. The sender sends the encrypted email and the encrypted symmetric key to the recipient.
    e. The recipient decrypts the encrypted symmetric key with the recipient's private key.
    f. The recipient decrypts the email with the symmetric key.
- **Encrypting the actual data with the public key.** This isn't done as often but can be done with small messages.
    a. The sender encrypts the email with the recipient's public key.
    b. The sender sends the encrypted email to the recipient.
    c. The recipient decrypts the email with the recipient's private key.

It's also worth comparing the keys used to encrypt data and the keys used to encrypt a digital signature:

- The sender's private key is used to encrypt the digital signature.
  In contrast, the recipient's public key is used to encrypt email content.
- The sender's public key is used to decrypt the digital signature.
  In contrast, the recipient's private key is used to decrypt email.

Can you answer these questions?

You can find the answers to these questions at the end of the chapter.

1. What benefits are provided from S/MIME?
2. What does PGP use instead of a PKI to distribute certificates?

Objective 5.10: Understand Public Key Infrastructure (PKI)

A Public Key Infrastructure (PKI) includes multiple elements to support the creation, distribution, and management of certificates. Just as the TCP/IP suite of protocols has a lot of a depth, a PKI also has a lot of depth. However, there are some basic components that are very important to understand when preparing for the CISSP exam.

Exam need to know...

- PKI components
  For example: What is the difference between a certificate authority and a registration authority? What determines whether a certificate is trusted?

PKI

A PKI includes all the elements necessary to support the creation, distribution, and management of certificates. A certificate authority (CA) issues and manages certificates. A registration authority (RA) is sometimes used and accepts requests for certificates, verifies requestor's information, and forwards the information to the CA.

True or false? A registration authority (RA) provides the key for a certificate and signs it.

Answer: *False.* A CA signs the certificates. The RA (when used) performs only registration activities such as verifying the requestor but does not issue certificates or provide keys. The requesting entity creates the key pair, provides the public key for the certificate, and keeps the private key private.

Many public CAs, such as VeriSign, have root certificates installed in an operating system's trusted root certification authority store by default. These CAs are trusted, and any certificate created by the CA is also trusted.

True or false? A key goal of public key cryptography is to enable two entities to share a symmetric key over a public network without previously coordinating with each other.

Answer: *True.* Public key cryptography allows two entities to privately share a symmetric key.

The PKI supports transmission of a certificate with the public key and is an integral element in privately sharing symmetric keys. For example, imagine two computers (a workstation and a server) need to create a secure session over a network. Symmetric encryption uses a single key, but both computers need to know the key. If the symmetric key is sent in clear text, anyone can see it and use it to decrypt the data in the secure session. Instead, the following steps are used.

If Computer A and Computer B need to communicate securely, Computer A can retrieve Computer B's certificate. The certificate can be retrieved from a trusted public repository without contacting Computer B directly, or the certificate can be retrieved from Computer B.

1. The workstation creates a symmetric key. At this point, only the workstation knows the key.
2. The workstation retrieves the server's certificate. This certificate includes the server's public key.
    a. The workstation might receive the certificate from the server or from a trusted public repository without contacting the server directly.
    b. The workstation uses the PKI to check the certificate to ensure that it is valid and hasn't been revoked.
3. The workstation encrypts the symmetric key with the server's public key.
4. The workstation sends the encrypted symmetric key to the server.
5. The server receives the encrypted symmetric key and decrypts it with its private key. At this point, both the workstation and the server have the symmetric key but it is not available to anyone else.

EXAM TIP A PKI includes all the elements to support asymmetric or public key cryptography, but the two aren't the same. Public key cryptography refers to encryption and decryption with public and private keys and is dependent on the infrastructure to manage certificates through their lifetime.

MORE INFO VeriSign has published the VeriSign Trust Network Certification Practice Statement document, which provides in-depth coverage of its practices. This document provides excellent insight into common best practices used with public CAs. You can access it here: *http://www.verisign.com/repository/CPSv3.8.4_final.pdf.*

CAs can also be privately created by an organization and used only by personnel within the organization. For example, an organization might create a private CA to create and manage certificates used for smart cards.

MORE INFO Microsoft includes Active Directory Certificate Services (AD CS) as a free service on Windows Server products. It is integrated within Active Directory and can be used to create private enterprise–level CAs. The following link provides access to a full range of articles and guides related to AD CS: *http://technet.microsoft.com /windowsserver/dd448615.aspx.*

True or false? Certificate trust chains are all hierarchical, starting with a root CA.

Answer: *False*. Most CAs are hierarchical, but this isn't the only trust model. For example, a cross-certification trust model allows two root CAs to trust each other.

A certificate trust chain refers to multiple levels of CAs and their trust relationships. The majority of CA trust chains are hierarchical. The root CA issues certificates to intermediate CAs, and these intermediate CAs can also issue certificates to other intermediate CAs. The lowest-level CA can issue certificates to users or computers. As long as the root CA is trusted, any certificates issued by any of the other CAs in the chain are also trusted.

Can you answer these questions?

You can find the answers to these questions at the end of the chapter.

1. What is the purpose of a registration authority?
2. What is the primary benefit provided by a PKI?
3. What is required for an entity to trust a certificate it receives?

Objective 5.11: Understand certificate related issues

Certificates are integral to the use of public key cryptography, and it's important to have a basic understanding of what is contained in a certificate. Additionally, certificates can be compromised, and if so, the CA needs to inform users that the certificate should no longer be trusted. This section covers basics on the contents of an X.509 certificate and methods used to inform users when a certificate has been revoked.

Exam need to know...

- Certificates
  For example: What is the standard used to define a certificate? What is the difference between a Class 1 certificate and a Class 3 certificate?
- Validating certificates
  For example: How are revoked certificates published? What are the two methods that a client can use to check the validity of a certificate?

Certificates

The standard for certificates is the X.509 standard defined by the International Telecommunications Union (ITU). The primary certificate in use is the X.509 v3 certificate. Certificates commonly have a serial number, information about the CA that issued the certificate, validity dates, who it was issued to, and the public key.

True or false? A Class 1 certificate provides the highest assurance level of an organization's authenticity.

Answer: *False*. A Class 1 X.509 certificate provides the lowest assurance level. A Class 3 certificate (including a Class 3 certificate with Extended Validation) provides the highest level of assurance. It requires additional steps in the registration process, such as in-person registration or registration using notarized credentials.

> **MORE INFO** VeriSign has published the VeriSign Trust Network Certification Practice Statement document, which provides in-depth coverage of its practices. This document provides excellent insight into common best practices. You can access it here: *http://www.verisign.com/repository/CPSv3.8.4_final.pdf*. RFC 3280 documents the profiles and use of X.509 v3 and v2 certificates (*http://www.ietf.org/rfc/rfc3280*).

Certificates have multiple purposes. Some of the common purposes include the following:

- Encrypt email messages
- Digitally sign email messages
- Digitally sign software
- Detect modifications in digitally signed software
- Validate the identity of remote computers
- Identify individuals (for example, by embedding a certificate in a smart card)

Validating certificates

If a certificate or its associated private key has become compromised, the CA will revoke the certificate. Certificate users first check the validity dates to ensure that the certificate is not expired and then query the issuing CA to determine whether a certificate is valid.

True or false? A CA will publish a certificate revocation list (CRL) by using an X.509 v2 certificate.

Answer: *True*. A CA publishes a CRL as an X.509 v2 certificate. The CRL includes the serial numbers of all certificates that have been revoked along with a reason why the certificate was revoked.

A CA might revoke a certificate for several reasons, such as the following:

- The private key is compromised.
- The CA is compromised.
- The certificate is put on hold.

True or false? The Online Certificate Status Protocol (OCSP) is used by clients to validate a certificate.

Answer: *True*. Instead of requesting a copy of the CRL, systems can check the status of a certificate by using OSCP.

With OCSP, the client system sends the serial number to the CA with a request for its status and the CA answers with a status of healthy, not healthy, or unknown. Healthy indicates the certificate is valid, and unhealthy indicates the certificate has been revoked and should not be trusted. A status of unknown indicates that the CA doesn't have a record of the certificate's serial number.

> **EXAM TIP** The two primary methods of checking a certificate are by retrieving the CRL or using OCSP. The CRL is published as an X.509 v2 certificate, but the list of revoked certificates can become extensive. Using OCSP requires less bandwidth.

> **MORE INFO** RFC 2560 documents OCSP; it is available here: *http://www.ietf.org/rfc/rfc2560.txt*.

Can you answer these questions?
You can find the answers to these questions at the end of the chapter.

1. What are the two methods used to validate a certificate?
2. What is added to the CRL to identify revoked certificates?
3. Who maintains the CRL?

Objective 5.12: Understand information hiding alternatives (e.g., steganography, watermarking)

Steganography and watermarking are two additional methods included in cryptography topics. In general, these methods are referred to as a type of security through obscurity. Messages aren't encrypted but instead are hidden so that casual users cannot detect their presence but knowledgeable people can easily detect them.

Exam need to know...
- Information hiding alternatives
  For example: What types of files are typically used with steganography? How can steganography be detected?

Information hiding alternatives

Steganography refers to hiding data within data. Messages can be embedded within a graphic file, an audio or video file, a text-based document, or even a blog or message posted on an Internet site. For example, the first letter of every fifth word in a blog post can be used to transmit a message.

True or false? It's possible to include a message within a graphic file by modifying the most significant bits of several bytes.

Answer: *False*. Steganography allows someone to embed a message within a graphic file by modifying the least significant bits of a file, not the most significant bits.

Data is stored in bytes. By modifying the least significant bit of a byte, it's possible to make changes that are barely perceptible to the casual observer. However, they can be used to leave extensive messages if someone knows where to look for the change. In a picture file, the change might make a very slight change in a red, green, or blue color. In an audio or video file, the change might make a very slight change in the sound pitch or video display.

For example, the decimal number 255 is designated in binary by the following bits: 1111 1111. Modifying the least significant bit (the last 1) of 255 decimal changes this to 1111 1110, or 254 decimal. If the most significant bit (the first 1) is modified, it becomes 0111 1111, or 127 decimal. The difference between 255 and 254 is minor and not noticeable, but the difference between 255 and 127 is significant.

The three elements of steganography are as follows:

- The carrier, which is the file that has the hidden message.
- The medium, which is the specific type of file, such as an .mp3 or .png file. This is sometimes called the stego-medium.
- The payload, which is the hidden message.

EXAM TIP Hashing algorithms are useful to detect steganography methods. By regularly creating a hash of website images and comparing them to previously captured hashes, it's possible to detect when any of the images have been modified. The images will likely look the same to any viewer, but the hashes will be significantly different if a message has been embedded within the file.

MORE INFO While there are many tools available to use and detect steganography, there is little proof that it is being done on the Internet. The following page provides a comprehensive listing of different steganography tools currently available: *http://www.jjtc.com/Security/stegtools.htm.*

Watermarking is the use of additional information in a file or a document. For example, it's common to use watermarks on checks to deter check fraud. Watermarks might be clearly visible to anyone looking for them or visible using only special methods.

Many checks have clearly visible watermarks on the back, using repeated words such as "Original Document." These are not easily copied by someone trying to re-create the checks. On most United States currency, a watermark is embedded in the paper by using an image of the fully visible portrait. This can be viewed on the right edge of the bill when held up to the light but is otherwise not visible.

Can you answer these questions?

You can find the answers to these questions at the end of the chapter.

1. What is it called when a message is embedded in file?
2. What bits are modified when messages are embedded in a file?
3. What cryptographic method is used to identify files that might have embedded messages?

Answers

This section contains the answers to the "Can You Answer These Questions?" sections in this chapter.

Objective 5.1: Understand the application and use of cryptography.

1. Data at rest is any data stored on media. This includes data stored as a file on internal or external disks, optical discs, tapes, or USB flash drives.

2. Encryption methods are used to protect against the loss of confidentiality for any type of data, including data at rest and data in transit.
3. SSH is used to encrypt FTP traffic (as SFTP). SSH is also used to encrypt many other protocols.

Objective 5.2: Understand the cryptographic lifecycle (e.g., cryptographic limitations, algorithm/protocol governance)

1. Cryptographic keys are longer, resulting in strong cryptographic security. The longer keys in use today weren't possible with many computers in use 10 years ago. Similarly, 10 years from now, today's cryptographic keys will likely be trivial in comparison to the keys in use then.
2. NIST publishes special publications (in the SP 800 series) and Federal Information Processing Standards (FIPS) documents. Both are freely available from the NIST research site.
3. Publications from NIST are published by the United States government and are freely available. ISO publications are international standards, and many must be purchased.

Objective 5.3: Understand encryption concepts

1. Encryption uses an algorithm and a key, and longer keys result in more complexity of the encrypted data. For example, data encrypted with a 256-bit key is significantly harder for an attacker to decrypt than data encrypted with a 56-bit key.
2. Symmetric cryptography uses a single key to encrypt and decrypt a piece of data.
3. The private key is never shared when using asymmetric cryptography. In contrast, the public key is commonly shared. A public key is embedded in a certificate, and the certificate is passed to other entities for use in public key cryptography.
4. The server's private key is used to decrypt the session key in an HTTPS session. The client creates the session key, encrypts it with the server's public key, and sends it in an encrypted format to the server.
5. SHA1 creates 160-bit message digests (or hashes).
6. A hashing algorithm should always create the same hash when executed against the same piece of data. If the hashes are different, you know that the original data has changed. That is, you know that the data has lost integrity.

Objective 5.4: Understand key management processes

1. Longer cryptographic keys should be used to protect more valuable data. For example, AES-128 (with 128-bit keys) might be sufficient to encrypt some data, but an organization might choose to use AES-256 (with 256-bit keys) to encrypt more valuable data.

2. The lifetime of a cryptographic key should be shorter for more valuable data. That is, keys should be changed more often as the value of the data increases.
3. Key recovery methods are designed to recover either the key or the data encrypted with the key. If the original key is lost but a copy of the key can be recovered, this recovered key can be used to decrypt the data. In some recovery methods, a designated recovery agent has an alternate key that can be used in place of the original key.
4. Private keys (which are matched to public keys) are held in escrow. Additionally, symmetric keys can also be held in escrow. There is no need to keep public keys in escrow.

Objective 5.5: Understand digital signatures

1. A digital signature provides authentication of the sender, integrity of the message, and non-repudiation.
2. The sender's public key is needed to verify a digitally signed message. A digital signature is created by encrypting the hash of a message with the sender's private key, and only the sender's public key can decrypt it. If the sender's public key decrypts the digital signature, it verifies that it was signed with the sender's private key.
3. A digital signature is created by encrypting a hash of the message with the sender's private key.
4. Confidentiality is provided by encryption, but a digital signature does not encrypt the message and does not protect the confidentiality of the message. If only a digital signature is used, the hash of the message is encrypted but the message isn't encrypted. If confidentiality of the message is important, the message should be encrypted, and it is possible to encrypt the message and include a digital signature.

Objective 5.6: Understand non-repudiation

1. The digital signature is encrypted with the sender's private key. If the sender's public key can decrypt it, it must have been encrypted with the sender's matching private key.
2. A key requirement for non-repudiation from an audit trail is a strong, secure authentication method. If users use poor password security, such as writing their passwords down or sharing their passwords, other users can log on to their accounts. Actions by these other users will be recorded, but it won't accurately reflect who took the actions.

Objective 5.7: Understand methods of cryptanalytic attacks

1. A salt is a random set of bits added to a hash. It defeats rainbow tables by forcing them to be significantly larger to accommodate the additional bits.

2. Changing encryption keys often reduces the success of a ciphertext-only attack. Each time the key is changed, the attack has to start over to decrypt the data with the new key.
   3. Implementation attacks commonly analyze environmental factors such as power, timing, and clock signals. Differential analysis provides insight into the keys.

Objective 5.8: Use cryptography to maintain network security

   1. Link encryption encrypts the data and the routing information. If intercepted, the attacker can see the routing information between two routers but not the original source and destination addresses. In contrast, end-to-end encryption encrypts only the data.
   2. The client creates the symmetric key within an HTTPS session. This key is also known as the session key.
   3. The client encrypts the symmetric key with the server's public key. This encrypted session key is then sent over the network. It can be decrypted only with the server's matching private key, and only the server has access to this private key.
   4. A symmetric key is used to encrypt data within an HTTPS session. The server's public and private keys are used only to privately share the symmetric key.
   5. A security association identifies the different cryptographic protocols that a secure IPsec channel will use and also includes the keys used in this channel.

Objective 5.9: Use cryptography to maintain application security

   1. S/MIME can be used to encrypt email and create digital signatures. It uses public and private keys and is dependent on a PKI to transfer certificates.
   2. PGP was originally designed to use a web of trust instead of a PKI.

Objective 5.10: Understand Public Key Infrastructure (PKI)

   1. A registration authority collects and validates registration information to relieve some of the workload of the CA. It does not issue or manage certificates.
   2. The primary benefit provided by a PKI is that it allows entities to privately share symmetric keys with each other without any previous coordination. The PKI supports the transfer of certificates over a public network, which includes a public key used with a matching private key for encryption and decryption. The PKI also provides validation of the certificates so that entities know that the contents of the certificate can be trusted.
   3. The entity must trust a CA in the CA trust chain. For example, if the entity trusts the root CA, it trusts all certificates issued by any CAs in the trust chain.

Objective 5.11: Understand certificate related issues

1. The two methods used to validate a certificate are by checking a copy of the CRL or by using OCSP.
2. The CRL is an X.509 v2 certificate that includes a list of serial numbers of revoked certificates. The serial number is assigned to the certificate when it is created and is used to identify it.
3. The CA maintains the CRL. When the CA revokes a certificate, it updates the CRL and makes it available to any entities that request a copy.

Objective 5.12: Understand information hiding alternatives (e.g., steganography, watermarking)

1. Steganography is the practice of embedding a message in a file.
2. When steganography is used to embed messages within a file, the least significant bit of various bytes is modified. These changes are relatively small and not detectable in a typical image, audio, or video file.
3. Hashing algorithms are used to detect changes in files. Even though a file won't have a change that is detectable by looking at or listening to the file, the resultant hash is significantly different if even a single bit is changed.

CHAPTER 6

Security architecture & design

The Security Architecture & Design domain builds on the core security principles of confidentiality, integrity, and availability. The focus is on the models, structures, and standards that are used to design, implement, monitor, and secure information technology (IT) systems and networks.

This chapter covers the following objectives:

- Objective 6.1: Understand the fundamental concepts of security models (e.g., Confidentiality, Integrity, and Multilevel Models)
- Objective 6.2: Understand the components of information systems security evaluation models
- Objective 6.3: Understand security capabilities of information systems (e.g., memory protection, virtualization, trusted platform module)
- Objective 6.4: Understand the vulnerabilities of security architectures
- Objective 6.5: Understand software and system vulnerabilities and threats
- Objective 6.6: Understand countermeasure principles (e.g., defense in depth)

Objective 6.1: Understand the fundamental concepts of security models (e.g., Confidentiality, Integrity, and Multi-level Models)

Several different security models are used to enforce the basic security concepts of confidentiality and integrity. Some common models that are frequently mentioned are the Bell-LaPadula, Biba, the Chinese Wall, and Clark-Wilson models. Ideally, all users accessing a system will have the same clearance levels, be approved for the same level of data classifications, and have a need to know for all of the data on the system. In reality, most systems are multilevel and have a mixture of users accessing the system.

Exam need to know...

- Security models
  For example: What is the primary difference between the Bell-LaPadula model and the Biba model? What are the three elements in the Clark-Wilson model?

Security models

The Bell-LaPadula is a mandatory access control (MAC) model that has a primary focus of ensuring confidentiality. It helps prevent unauthorized individuals from accessing data.

True or false? The Bell-LaPadula model uses a simple security property rule of no read-up.

Answer: *True*. The simple security property rule (no read-up) ensures that subjects cannot read data at a higher classification, which helps enforce confidentiality.

> **NOTE** Security models commonly use the terms *subject* and *object*. A subject is any entity (including a user) that can access a resource such as a file or folder. An object is any type of resource that can be accessed by a subject.

The Bell-LaPadula model includes the following three rules:

- **The Simple Security Rule** Commonly stated as no read-up, this prevents access to data at higher classification levels. This rule directly addresses confidentiality.
- **The *-property (star property) rule** Commonly stated as no write-down, this prevents an entity from modifying data at lower classifications. This rule addresses integrity.
- **The Discretionary Security property rule** An access control matrix is used to specify permissions for specific resources.

The Biba model has a primary focus of enforcing integrity. It includes the following rules:

- **The *-Integrity Axiom (star Integrity Axiom)** Commonly stated as no write-up, this rule prevents an entity from modifying data at higher classifications. This rule directly addresses integrity.
- **The Simple Integrity Axiom** Commonly stated as no read-down, this rule states that a subject granted access at a specific level should not be able to read objects at a lower level.
- **The Invocation property rule** This rule isn't always mentioned, but it indicates that a subject cannot invoke or call upon a subject at a higher integrity level. That is, a subject cannot send a message or a request for a service to a subject at a higher level.

> **EXAM TIP** The Bell-LaPadula model uses a no read-up and a no write-down rule, with a primary focus on confidentiality. In contrast, the Biba model uses a no write-up and a no read-down rule, with a primary focus on integrity.

MORE INFO Chapter 1, "Access Control," briefly mentions the Bell-LaPadula and Biba models in the context of MAC models. Chapter 3, "Information Security Governance & Risk Management," directly addresses the goals of confidentiality, integrity, and availability as the CIA triad (or the AIC triad). As a reminder, confidentiality prevents disclosure of information to unauthorized individuals, integrity ensures that unauthorized modifications are discovered, and availability ensures that systems are operational when they are needed.

True or false? The Clark-Wilson access model provides integrity by forcing subjects to access objects through an application.

Answer: *True*. The Clark-Wilson access control model controls access to data through an application.

Access to databases is often controlled by using the Clark-Wilson access control model. For example, a database could have an Employee table with first name, last name, address, phone number, and salary columns. The application controls access to these columns based on the permissions of the user. Some users would see all the columns except the salary data, while other users with appropriate permissions would see all the columns.

Similarly, databases often use database views. A view provides a limited view of the table by including only specific columns. For example, a view of the same Employee table could include the first name, last name, address, and phone number columns. Users with access to the view (but not the underlying table) would not have access to the salary data because they can access only the columns in the view.

EXAM TIP The Clark-Wilson model uses an access triple: the subject (such as a user), the object (the underlying data), and an application. The only way a subject can access the object is through the application. An additional benefit is that the application has full control of the entire transaction to help maintain consistency of the data.

True or false? A primary purpose of the Brewer and Nash model is to reduce conflict-of-interest situations.

Answer: *True*. To prevent a conflict of interest, the Brewer and Nash model (also called the Chinese Wall model) is designed to limit the amount of information available to subjects and objects.

EXAM TIP The Chinese Wall model is commonly implemented in financial institutions. The goal is to separate the information available within the corporate executive group from those giving advice to outside clients. That is, brokers giving information to customers are shielded from inside information that could affect the advice they give to clients.

True or false? A positive security model focuses on what is allowed rather than what is not allowed.

Answer: *True*. Positive security models are effective at preventing new and unknown attacks. For example, many firewalls have rules identifying what traffic is allowed, and all other traffic is blocked.

Positive security models are also referred to as *whitelisting*. A website whitelist can include a list of website addresses that are allowed and block all others. Similarly, a whitelist can include a list of known email addresses and block all others.

> **EXAM TIP** Firewalls commonly include a deny-all rule placed at the end of the access control list as part of a positive security model. This denies all traffic that has not been previously allowed. Some environments use whitelisting to block malware and authorize only specific applications. All non-authorized applications are blocked.

True or false? A discretionary access control model includes multiple access controls, such as read, write, and execute.

Answer: *True*. Discretionary access control models include multiple permissions, including read, write, and execute. These permissions are commonly included in an access control list for an object.

True or false? In a system high-security mode, all users have proper clearance for the entire system, have formal access approval for the entire system, and have a valid need to know all the data stored on the system.

Answer: *False*. This defines dedicated security mode, not system high-security mode. In system high-security mode, users have a need for some of the data on the system but not all of it.

Common security modes include the following:

- **Dedicated security mode** All personnel are cleared to access the system, have formal access approval for all of the data on the system, and have a valid need to know all of the data.
- **System high-security mode** All personnel are cleared to access the system, have formal access approval for all of the data, and have a need to know *some* of the data.
- **Compartmented security mode** All personnel are cleared to access the system, have formal access approval for *some* of the data, and have a need to know *some* of the data.
- **Multilevel security mode** All personnel are cleared to access *part* of the system, have formal access approval for *some* of the data, and have a need to know *some* of the data.

> **EXAM TIP** It is not common to have a system running in dedicated security mode. Instead, additional security measures are needed to control access to the system based on a valid need to know, clearances, and access approval based on job responsibilities.

> **MORE INFO** The United States Department of Defense document 5200.28 includes a list of many security terms and includes definitions for security modes. You can access it here: *http://csrc.nist.gov/groups/SMA/fasp/documents/c&a/DLABSP/d520028p.pdf*.

Can you answer these questions?

You can find the answers to these questions at the end of the chapter.

1. What is included in the Clark-Wilson triple?
2. A firewall has a rule at the end that blocks all traffic that hasn't been previously allowed. What type of security model is this?
3. What must be met for a system to operate in dedicated security mode?

Objective 6.2: Understand the components of information systems security evaluation models

Several different evaluation models are available to help an organization evaluate security related to different products and services. The Common Criteria is an international standard used to evaluate a wide variety of products and systems. In contrast, some models are used only in specific instances. For example, the Payment Card Industry Data Security Standard (PCI DSS) is used by credit card companies to help ensure that retailers protect credit card data.

Exam need to know...

- Product evaluation models (e.g., Common Criteria)
  For example: How many Evaluation Assurance levels are in the Common Criteria? Which Evaluation Assurance Level is the most secure?
- Industry and international security implementation guidelines (e.g., PCI DSS, ISO)
  For example: What is the primary purpose of PCI DSS? What are the primary ISO documents used for IT security management?

Product evaluation models (e.g., Common Criteria)

Product evaluation models are designed to evaluate systems for security. The Common Criteria is an international standard (also known as ISO/IEC 15408). It is used by many organizations to compare different products against each other for specific security protection.

True or false? The Common Criteria is a fixed security model that does not change.

Answer: *False.* The Common Criteria is intended to evolve to meet the needs of newer systems. In contrast, the Trusted Computer System Evaluation Criteria (TCSEC, or the Orange Book) standard previously used by the United States government became obsolete, and the Common Criteria was adopted in 2005 in its place.

The Common Criteria defines security controls by using security functional requirements and security assurance requirements. Multiple levels are used to define various levels of security.

True or false? Evaluation Assurance Level (EAL) 3 defines a system as having been methodically tested and checked.

Answer: *True*. EAL 3 is one of seven EAL assurance levels, and it indicates the system has been methodically tested and checked.

The following list identifies all seven levels, from the highest assurance levels to the lowest:

- **EAL 7**   The system has a formally verified design, and it has been tested.
- **EAL 6**   The system has a semiformally verified design, and it has been tested.
- **EAL 5**   The system was semiformally designed, and it has been tested.
- **EAL 4**   The system has been methodically designed, tested, and reviewed.
- **EAL 3**   The system has been methodically designed, tested, and checked.
- **EAL 2**   The system has been structurally tested.
- **EAL 1**   The system has been functionally tested.

True or false? A Security Target (ST) is a statement of security needs for a target of evaluation (TOE).

Answer: *True*. The ST identifies the goal in terms of security requirements or security needs. The TOE is the software, firmware, hardware, or system that is being evaluated and must meet the requirements specified in the ST.

The following list identifies some common terms used within the Common Criteria:

- **TOE**   The software, firmware, hardware, or system that is being evaluated.
- **Protection Profile (PP)**   A document that identifies security requirements needed for different purposes.
- **Security Functional Requirements**   A list of functions that a product must provide.
- **Security Assurance Requirements**   A list of metrics used to verify that the product meets the specific functional requirements.

EXAM TIP   The Common Criteria is hundreds of pages contained in several parts, but you aren't expected to know all the details. However, you should be aware that it is a common standard and that it includes security functional requirements and security assurance requirements, and you should be aware of the different EALs.

MORE INFO   You can download Common Criteria documents from the Common Criteria portal here: *http://www.commoncriteriaportal.org/cc/*. Part 1 is the introduction and an outline of the general model. Part 2 identifies the security functional requirements, and Part 3 identifies the security assurance requirements.

Industry and international security implementation guidelines (e.g., PCI DSS, ISO)

Several organizations have been created to provide specific guidance for organizations related to IT security and other topics.

True or false? The primary purpose of the PCI DSS is to protect consumers.

Answer: *False*. The PCI DSS is a standard created by several credit card companies with the goal of reducing fraud. Consumers benefit, but the companies and the retailers often suffer the majority of the losses from credit card fraud. Usually, consumers can reverse fraudulent charges by reporting them within 30 days of receiving a bill.

The PCI DSS includes the following twelve specific requirements based on six principles. Retailers using credit cards are responsible for implementing these requirements and principles to protect their network and credit card data.

- **Build and Maintain a Secure Network**

    Requirement 1: Install and maintain a firewall.

    Requirement 2: Do not use defaults, such as default passwords.

- **Protect Cardholder Data**

    Requirement 3: Protect stored data.

    Requirement 4: Encrypt transmissions.

- **Maintain a Vulnerability Management Program**

    Requirement 5: Use and update antivirus software.

    Requirement 6: Develop and maintain secure systems.

- **Implement Strong Access Control Measures**

    Requirement 7: Restrict access to data.

    Requirement 8: Use unique logons for each user. Don't share user names and passwords.

    Requirement 9: Restrict physical access.

- **Regularly Monitor and Test Networks**

    Requirement 10: Track and monitor all access to systems and data.

    Requirement 11: Regularly test security.

- **Maintain an Information Security Policy**

    Requirement 12: Maintain a security policy.

EXAM TIP Most assessments under the PCI DSS are done internally by retailers as self-assessments. However, after an incident, retailers are often required to hire an outside Qualified Security Assessor that has been trained by PCI DSS to perform a third-party assessment and ensure PCI DSS compliance.

MORE INFO The PCI DSS can be downloaded here: *https://www.pcisecuritystandards.org/documents/pci_dss_v2.pdf*. It provides detailed information about all the requirements and principles.

True or false? ISO/IEC 27002 provides best practice recommendations for IT security.

Answer: *True.* ISO/IEC 27002, "Information technology - Security techniques - Code of practice for information security management," provides best practice recommendations related to IT security.

> **EXAM TIP** ISO/IEC 27001, "Information technology – Security techniques – Information security management systems – Requirements," and ISO/IEC 27002 combined provide common guidelines that many organizations implement for IT security. These documents were previously known as BS7799 and ISO/IEC 17799, but they were renumbered to coincide with other numbering in the ISO/IEC series.

> **NOTE** ISO is not an acronym for the International Organization of Standardization. Instead, it's short for *isos*, which is a Greek word for *equal*. Many standards are published as ISO and International Electrotechnical Commission (IEC) standards (as ISO/IEC). You can read their definition here: *http://www.iso.org/iso/home/about.htm*.

True or false? The Information Technology Infrastructure Library (ITIL) defines five phases within IT management.

Answer: *True.* ITIL includes five books, and each book describes one of the five phases of ITIL.

The five phases described within ITIL are as follows:

- **Service Strategy** In this phase, services are evaluated to determine whether they have sufficient value to the organization.
- **Service Design** In this phase, services are designed for ultimate implementation within the organization.
- **Service Transition** This phase includes adding new services, modifying existing services, or removing obsolete services. A primary goal is to ensure that the transition does not adversely affect existing services, and this phase is commonly referred to as *change management*.
- **Service Operation** In this phase, daily operations and support of any service are handled.
- **Continual Service Improvement** This phase focuses on measuring and monitoring services and processes to determine areas where services can be improved.

> **MORE INFO** You can access more information about ITIL here: *http://www.itil-officialsite.com/*.

Can you answer these questions?

You can find the answers to these questions at the end of the chapter.

1. How many EALs are in the Common Criteria?
2. When using Common Criteria, what identifies the security requirements for a system?
3. How many principles and requirements are included in the PCI DSS?

Objective 6.3: Understand security capabilities of information systems (e.g., memory protection, virtualization, Trusted Platform Module)

Security for individual systems is part of an overall defense in depth strategy. In this context, individual systems are end user systems or servers, and security at this level helps to protect data stored on the systems from compromise. This objective also addresses some security concerns related to virtualization.

Exam need to know...

- Security capabilities of information systems
  For example: What are concerns related to data in use? What is provided by a trusted platform module?

Security capabilities of information systems

When protecting data, the two primary focuses are data at rest (stored as files on media) and data in transit (sent over a network). However, a third focus is related to data in use.

True or false? Data in use includes any data in volatile memory.

Answer: *True*. Data in use is stored in volatile memory and is used by the operating system or applications.

Applications should automatically deallocate memory that is no longer needed. This prevents memory leaks, which can consume progressively more and more memory and ultimately crash a system. It also ensures that sensitive data is not left unprotected in memory. Some applications include garbage collection routines that periodically sanitize areas of memory to ensure that these unused variables are no longer readable.

True or false? Virtual memory is stored on a hard disk as a file.

Answer: *True*. Virtual memory uses disk space to extend the size of the physical memory.

The virtual memory file (also called a swap file or a paging file) can contain sensitive information that would normally be deleted from a system's volatile memory when the system is powered off. However, because it is stored on the drive, it can be retrieved and examined if an attacker has access to it.

> **EXAM TIP** Random access memory (RAM) is volatile and lost if the system is turned off, but valuable information remains while the system is powered on. Data stored in temporary variables in clear text can be accessed by anyone with physical access to the system, and some of this data is stored in a file on the hard drive. Forensics tools are available to retrieve this information rather easily, and these tools are also available to attackers.

True or false? Virtualization allows an organization to reduce its carbon footprint and save money related to server space, power, and air conditioning.

Answer: *True*. Multiple virtual systems can be hosted on a single physical server, reducing the number of servers required. This also reduces power consumption and heating ventilation air conditioning (HVAC) requirements.

Virtualization techniques allow a single physical system to host multiple instances of virtual machines (VMs). Virtualization is becoming more and more popular due to the many benefits it provides to an organization.

A VM is a virtual instance of a fully functional operating system. When the VM is turned off, it is stored as one or more files on a computer and can easily be exported from the current host system and imported to another host system.

> **MORE INFO** Microsoft systems have supported Hyper-V and virtualizations for many years. The following page provides an introduction to server virtualization and also includes links to multiple resources on virtualization: *http://www.microsoft.com/en-us/server-cloud/windows-server/server-virtualization.aspx*.

True or false? A successful VM escape attack allows the attacker to access the host system via one of the virtual systems.

Answer: *True*. In some cases, an attacker can run code within a VM and access the host operating system. When successful, the attacker has full control of the host and all of the VMs running on the host.

> **EXAM TIP** An important protection against VM escape attacks is to keep all systems up to date, including both physical and virtual systems. Virtual systems need all of the same updates as physical systems, and if they are not kept up to date, they become vulnerable to potential attacks.

> **MORE INFO** You can see a formal definition of a VM escape here: *http://whatis.techtarget.com/definition/virtual-machine-escape*. The following article documents one of the first instances of a VM escape vulnerability in VMware software: *http://www.coresecurity.com/content/advisory-vmware*. A white paper hosted on NIST's site, titled "VMware Security Briefing," provides an overview of virtualization along with some best practices to prevent VM escape attacks: *http://nvd.nist.gov/scap/docs/2008-conf-presentations/day2/VMware_Security_NIST.pdf*.

True or false? A Trusted Platform Module (TPM) helps protect data on laptop disk drives from loss of confidentiality.

Answer: *True*. TPMs are often included in laptop systems, and when enabled, they can be used to encrypt the hard drive. The encryption protects the drive from being accessed by unauthorized individuals.

A TPM is a chip on a motherboard used for encryption of entire disk volumes.

> **MORE INFO** BitLocker Drive Encryption is a feature on many editions of Microsoft operating systems for laptops, desktops, and servers. The following page provides an overview of how this is used with a TPM: *http://technet.microsoft.com/library/cc732774.aspx*.

Can you answer these questions?

You can find the answers to these questions at the end of the chapter.

1. What can an operating system or an application do to protect data in use?
2. What is the difference in update requirements for systems running on physical servers and systems running as VMs?
3. What can be used to encrypt a disk drive volume?

Objective 6.4: Understand the vulnerabilities of security architectures

Although security architectures have steadily improved over the years, vulnerabilities still exist. This objective covers some common methods used to reduce the associated risks due to these vulnerabilities.

Exam need to know...

- System (e.g., covert channels, state attacks, emanations)
  For example: What is the purpose of an operating system security kernel? What is a covert channel?
- Technology and process integration (e.g., single point of failure, service-oriented architecture)
  For example: How are single points of failure eliminated? What is service-oriented architecture?

System (e.g., covert channels, state attacks, emanations)

System security architecture refers to the different elements within an operating system and the physical system that together help provide system security. Within an operating system, the security kernel provides important protection mechanisms.

True or false? A security kernel is a group of internal security processes designed to monitor and control input and output operations, memory management, and process execution.

Answer: *True*. The security kernel and its associated reference monitor are responsible for monitoring all input, output, memory access, and process execution. The security kernel can be hardware, software, or firmware. It is the core of an operating system responsible for security.

Operating systems are built-in layers, with the security kernel at the core. Operating system utilities and applications are built on top of the operating system at the outer core. The security kernel uses a reference monitor to control access by subjects to any authorized objects based on access control lists. The layers are sometimes referred to as rings, with Ring 0 listed as the inner core where the security Kernel is located.

EXAM TIP The security kernel is the inner ring or core of the operating system. It interacts with the reference monitor to control access to objects by subjects.

True or false? Covert channels commonly use tunneling to hide unauthorized traffic.

Answer: *True*. A covert channel uses tunneling so that one unauthorized protocol is encapsulated within another authorized protocol.

Covert channel attacks masquerade as legitimate traffic to bypass security controls. After they bypass a control such as a firewall, the encapsulated protocol can launch the attack undetected.

EXAM TIP Covert channels are used in many different types of attacks. Bot herders have been known to control computers within a botnet by using Internet Relay Chat (IRC) as a covert channel.

MORE INFO Many attackers are seeking to exploit vulnerabilities in IPv6. The following article talks about how IPv6 can be used to create a covert channel by using ICMPv6 and 6to4 tunneling: *http://www.securityfocus.com/news/11406*.

True or false? A difference in the time of check (TOC) and time of use (TOU) of a user's permissions is an example of a race condition that can cause a vulnerability.

Answer: *True*. If there is a delay between the time when a user's permissions are checked (TOC) and when they are used (TOU), it can result in a software flaw or security vulnerability.

If a user's permissions are revoked between the time when the user's permissions were first checked and when they are used, the user can potentially have unauthorized access to data.

True or false? Data emanations can be blocked with a Faraday cage.

Answer: *True*. A Faraday cage is a shielded enclosure that can block radio frequency emanations from getting in or out.

Shielded enclosures can be large room-size enclosures to encompass a work space or small boxes to encompass a single system. A shielded enclosure prevents unauthorized individuals from receiving the data and can also block electromagnetic or radio frequency interference from garbling a signal.

Shielded twisted-pair (STP) cable prevents data from emanating from the cable and blocks interference. Fiber optic cable is more expensive, but data does not emanate from it and it is immune to interference.

> *EXAM TIP* The primary risk of data emanation is that unauthorized individuals might capture the data. Shielding methods are effective at blocking emanations. Additionally, cables carrying sensitive or classified data should be isolated from cables carrying unclassified data. If they are placed close to each other, it's possible for data in one cable to be picked up on the other cable. This is commonly known as *crosstalk*.

Technology and process integration (e.g., single point of failure, service-oriented architecture)

Even though systems are very complex, there are many technologies that streamline how they work together. Often these processes include redundancies to eliminate single points of failure.

> *MORE INFO* Chapter 7, "Operations Security," covers different system resilience and fault tolerance requirements. These are designed to eliminate single points of failure. For example, redundant array of independent disks (RAID) eliminates a disk drive as a single point of failure, and failover clusters eliminate a server as a single point of failure.

True or false? A service-oriented architecture (SOA) is a collection of services that can communicate and work together.

Answer: *True*. An SOA is a set of two or more services that can communicate with other. In this context, a service is a self-contained entity that can accept a request and provide a response.

For an example of SOA-based services, consider a purchase from an online retailer. During the checkout process, the customer provides credit card information. The retailer sends a service request to the appropriate credit card clearing house, receives a response, and finalizes the order. If the charge isn't approved, the retailer can simply prompt the user for another card. Similarly, after the purchase, a retailer can show the status of the shipment on its website. It sends a request to the shipper, the shipper sends the response, and the website displays the result.

> *MORE INFO* The article titled "Understanding Service-Oriented Architecture" on Microsoft's MSDN site provides a concise explanation of service-oriented architecture and shows how web services fit in. You can access it here: http://msdn.microsoft.com/en-us/library/aa480021.aspx.

Can you answer these questions?
You can find the answers to these questions at the end of the chapter.
1. Within an operating system, what is responsible for security?
2. What is a primary method used to prevent data emanations?
3. What is used to eliminate a single point of failure for disk drives?

Objective 6.5: Understand software and system vulnerabilities and threats

Software applications and systems have several specific vulnerabilities and threats that are important to understand. This objective requires you to have some basic knowledge about vulnerabilities from several different perspectives: the web, clients, servers, databases, and distributed systems.

Exam need to know...

- Web-based (e.g., XML, SAML, OWASP)
  For example: How can XML traffic be protected? What vendor sponsors OWASP?
- Client-based (e.g., applets)
  For example: What is the best protection against applets? What is whaling?
- Server-based (e.g., data flow control)
  For example: What is data loss protection? What is the difference between network-based data loss protection and an intrusion detection system?
- Database security (e.g., inference, aggregation, data mining, warehousing)
  For example: What is the difference between inference and aggregation? What is the effect of a deadlock on a database's performance?
- Distributed systems (e.g., cloud computing, grid computing, peer-to-peer)
  For example: What are the risks of storing data in the cloud? What are the risks from P2P software?

Web-based (e.g., XML, SAML, OWASP)

Web-based languages are based on the Extensible Markup Language (XML) and include the Security Assertion Markup Language (SAML).

True or false? One of the effective methods of preventing XML exploitation attacks is by encrypting the traffic with Secure Sockets Layer (SSL) or Transport Layer Security (TLS).

Answer: *True.* XML data sent over the Internet can be encrypted with SSL or TLS for confidentiality, integrity, and authentication. This is a primary protection against web service exploits. You can read about protecting web services here: *https://www.owasp.org/index.php/Web_Service_Security_Cheat_Sheet.*

The Open Web Application Security Project (OWASP) is a not-for-profit organization in the United States. They are vendor-neutral and have a wealth of information about safety and security of software.

> **MORE INFO** OWASP maintains a compressive list of "Cheat Sheets" that provide detailed information on protecting against a wide variety of different web-based attacks. You can access it here: *https://www.owasp.org/index.php/Cheat_Sheets*. XML and SAML are mentioned in Chapter 1 in the context of decentralized/distributed access control techniques. These are often used for single sign-on federated identities.

Client-based (e.g., applets)

The primary threat against client-based software is malicious software (malware) that runs on the client system. This can be in the form of scripted code, applets, ActiveX controls, and other executables. Malware can be included with webpages and as an attachment with email.

True or false? Signed applets provide assurances to users that an applet is safe.

Answer: *True*. As long as the applet is signed by a certification authority that is trusted, it provides a level of assurance to users that the applet is safe.

> **EXAM TIP** Both ActiveX controls and applets can be signed. When they are signed, a certificate identifies the original author of the ActiveX control or applet and provides a notification to the user if it has been modified since it was originally released.

True or false? Whaling is the practice of targeting executives with phishing emails.

Answer: *True*. Whaling will often target executives, such as chief executive officers, chief financial officers, or presidents.

Common vulnerabilities related to client email include the following:

- **Phishing** A phishing email goes to a mass number of users with the goal of tricking them into replying with personal information or clicking a malicious link. Sometimes the malicious link takes the user to a website that performs a drive-by download, installing malware on the user's system. Other times, the link takes the user to a bogus website that looks legitimate, with the goal of getting the user to enter their credentials.
- **Spear phishing** A spear phishing email targets a specific group of users with a phishing email. For example, it might target employees within a company with a spoofed From address that looks like it's coming from someone within the company.
- **Whaling** This is a phishing email that targets specific executives. It will often include information about the user, addressing the user directly by name, making the email appear more legitimate.
- **Beacons** Many emails include beacons that let the attacker know when the email has been opened. For example, a small graphic hosted on a website will be retrieved from the website when the graphic is displayed. This lets the attacker know that the email address is valid.

- **Malicious attachments** Attackers often include malicious attachments with spam, and when the user opens the attachment, it can infect the user's system. These can be obvious executable files or other types of files that support embedded code. For example, some Adobe Portable Document Files (PDFs) support embedded scripts and will run when the user opens the PDF.

EXAM TIP It's a common practice to prevent graphics from being displayed and to prevent email attachments from being automatically opened without additional action from the user. Additionally, filters at the boundary to the Internet, at email servers, and at email clients work together to detect and block malicious email and spam.

Server-based (e.g., data flow control)

One of the challenges with data being transmitted between systems is data leakage. Data leakage refers to the inadvertent or malicious disclosure of sensitive data to unauthorized personnel.

True or false? Data loss protection devices can detect whether users are sending sensitive data outside the organization.

Answer: *True*. Network-based data-loss prevention devices scan data being sent out of the network, looking for data that matches specific words or character strings. They can detect when users are sending sensitive data outside the organization.

For example, an organization might have a special project code named "Successful CISSP," which they have classified as confidential. They can then configure the data-loss prevention device to look for "Successful CISSP" and "confidential." If a character string is detected, the device blocks the transmission and sends an alert. Similarly, these devices can detect sensitive data, such as a Social Security number, by looking for a character string matching a specific format like this: ###-##-####.

EXAM TIP Data-loss prevention devices can be server-based (monitoring traffic leaving a server), networked-based (monitoring network traffic), and endpoint-based (monitoring user systems). Network-based data-loss prevention systems monitor traffic leaving the network. In contrast, intrusion detection and prevention systems monitor traffic coming into a network.

Database security (e.g., inference, aggregation, data mining, warehousing)

Databases can hold massive amounts of data. In some cases, this data is extremely valuable to the organization and if it is compromised, the losses can be catastrophic. Online transaction processing (OLTP) databases are used to capture transactions in real-time, and they are optimized for speed. Online analysis processing (OLAP) databases are used for analysis.

A data warehouse is a collection of data from multiple sources (often OLTP databases). Extract, transform, load (ETL) techniques are used to move data from other databases into OLAP databases while also ensuring consistency of the data. Data mining techniques can then extract actionable information from the OLAP databases used by decision makers.

True or false? Aggregation attacks occur when a user can access summary data but is not authorized to view details.

Answer: *True*. Aggregation in a database refers to performing queries on a group of data to get summary data, such as a total or an average. In an aggregation attack, the attacker can gain information about details by analyzing summary data.

For example, consider an organization that has two sales regions, named North and South. A user can run queries related to sales in the North region and run queries on total sales for the company, but the user is not authorized to run queries on sales in the South region. By running queries on the total sales and the North region, the user can identify sales for the South region.

True or false? Inference refers to the ability of a user to correctly guess (or infer) information by using access to other information.

Answer: *True*. An inference attack allows an individual to access data legitimately but learn additional information illegitimately. In the previous aggregation example, the user can use inference to determine sales in the South region.

> **EXAM TIP** Polyinstantiation is a common method used to prevent inference attacks in multilevel security databases. It allows the same data to be created as separate instances based on the security level of the data.

True or false? A database deadlock occurs when two or more resources are competing for access to the same data.

Answer: *True*. A deadlock condition stops access to data due to a data contention issue and can cause significant performance problems.

As an example of a deadlock, imagine that two threads are trying to access the Employee and Sales tables in a database. For example, thread 1 could lock the Employee table and then try to update the Sales table. At the same time, thread 2 could lock the Sales table and then try to update the Employee table. Neither thread will release its locks until it can complete its action on the second table, but neither thread can access the second table because it is locked by the other thread. Most database management systems will detect the deadlock by identifying one of the threads as a deadlock victim and terminating it. However, multiple deadlocks result in significant performance delays because each deadlock victim must repeat the thread.

> **EXAM TIP** Context-dependent and content-dependent access controls are programmed into an application to protect data. Context-dependent access controls prevent access to data based on the context of an operation. For example, a user must first complete steps A and B before access to certain data is allowed. Content-dependent access controls restrict access to data based on the user's permissions. For example, a program feature might be missing or dimmed if the user doesn't have the appropriate permissions to access it.

Distributed systems (e.g., cloud computing, grid computing, peer to peer)

Distributed systems refer to any system that is processing data from more than one location. Cloud computing is an example that has become extremely popular. It allows users to collaborate on projects from multiple locations as long as they have access to the Internet.

True or false? A primary risk of storing data in the cloud is the loss of confidentiality.

Answer: *True*. Data stored in the cloud is almost always stored by a third party, and the third party is responsible for control of this data.

If a third party hosting data in the cloud suffers a breach, it results in a loss of confidentiality for the original company or a loss of integrity if the attacker modifies the data. Additionally, a malicious employee at the third party's organization might be able to access the data. It's commonly recommended to encrypt all data that is stored in a third-party location.

> **EXAM TIP** Data stored in the cloud is at a high risk of being compromised. Encryption of the data provides protection, but the owner of the data should encrypt it. If it is encrypted at the cloud storage facility, personnel at this third party can potentially access the data.

True or false? The most important security concern related to peer-to-peer (P2P) computing systems is the violation of copyright laws.

Answer: *False*. While protecting intellectual property is important, P2P file sharing networks present two other risks that are more important to an organization—data leakage and malware.

Many versions of P2P file sharing software have the ability to share files from a user's system. Often this is the default configuration, and uneducated users are unaware of how to change the settings. Additionally, some versions can also share files from other systems in the network by using mapped drives. It's very common for files downloaded from P2P networks to be infected with malware, and unprotected systems are infected without the user's knowledge.

Can you answer these questions?

You can find the answers to these questions at the end of the chapter.

1. What is commonly used to protect XML data against loss of confidentiality?
2. What provides a user some assurance than an ActiveX control is safe?
3. What is the difference between a whaling attack and a spear-phishing attack?
4. What are common methods used within applications to protect data?
5. In relation to cloud computing, what is the best method of protecting sensitive data?

Objective 6.6: Understand countermeasure principles (e.g., defense in depth)

Security professionals know that implementing a single method of protection is never enough. Instead, countermeasures are needed to provide adequate protection.

Exam need to know...

- Defense in depth
  For example: What is defense in depth? How can you add defense in depth to a perimeter network?

Defense in depth

Defense in depth is a basic security principle that ensures that multiple layers of security are used. If a vulnerability exists or an attacker is able to bypass one security control, the overall security still limits or prevents a successful attack.

True or false? A perimeter network with two firewalls is an example of defense in depth for an internal network.

Answer: *True*. If an attacker is able to breach one firewall, the internal network is still protected with the second layer. This is often combined with defense diversity by using two different firewalls from two different vendors.

> **EXAM TIP** Defense in depth countermeasures can be applied in multiple layers of security. For example, firewalls are intended to control traffic, but intrusion detection and prevention systems can also be used to detect and block attacks. Similarly, a fence around a building provides one layer of physical protection, but multiple additional layers of security are included within the building, such as alarms, cameras, and cipher locks.

> **MORE INFO** Defense diversity is also discussed in Chapter 2, "Telecommunications and Network Security," in Objective 2.2 in the context of network access control devices.

Can you answer these questions?

You can find the answers to these questions at the end of the chapter.

1. What is the difference between defense in depth and defense diversity?

Answers

This section contains the answers to the "Can You Answer These Questions?" sections in this chapter.

Objective 6.1: Understand the fundamental concepts of security models (e.g., Confidentiality, Integrity, and Multi-level Models)

1. The Clark-Wilson triple includes the subject (such as a user), the object (the data being accessed), and an application. The application controls access to the object by the subject.
2. A system that identifies what is allowed and blocks everything else is a positive security model. It's sometimes referred to as whitelisting.
3. All personnel must be cleared to access the system, have formal access approval for all of the data contained in the system, and have a valid need to know all of the data in the system.

Objective 6.2: Understand the components of information systems security evaluation models

1. There are seven EALs in the Common Criteria. EAL 7 indicates the highest assurance level, and EAL 1 indicates the lowest assurance level.
2. The PP identifies the security requirements for a system. The system being evaluated is called the TOE.
3. The PCI DSS includes 6 principles and 12 requirements.

Objective 6.3: Understand security capabilities of information systems (e.g., memory protection, virtualization, trusted platform module)

1. Data in use refers to data stored in memory, and it can be protected with garbage collection routines to ensure that unused variables are removed from memory. When possible, sensitive data should not be stored in memory. For example, instead of storing a password in memory, the hash of the password should be stored. Physical security is also important to prevent anyone from accessing the system directly and using forensics tools to extract it.
2. There is no difference in the update requirements for physical systems and virtual machines. Both must be updated regularly.
3. A TPM can be used to encrypt a hard drive volume. This protects data on the disk even if the drive is stolen.

Objective 6.4: Understand the vulnerabilities of security architectures

1. The security kernel and its associated reference monitor are responsible for security within an operating system. They work together to ensure the subjects can access objects only when they have the appropriate permissions.
2. Shielding is commonly used to prevent data emanations. This includes shielding of enclosures and cables.
3. RAID subsystems are used to eliminate single points of failure for disk drives.

Objective 6.5: Understand software and system vulnerabilities and threats

1. SSL and TLS are commonly used to protect XML data.
2. When ActiveX controls are signed, it identifies the author and provides assurances that the ActiveX control has not been modified since it was released.
3. A whaling attack targets executives. A spear-phishing attack is a targeted attack against a group, but this group does not necessarily consist of executives.
4. Context-dependent access controls restrict access to data based on the context of an operation. Content-dependent access controls restrict access to data based on the user's permissions. Polyinstantiation is used to prevent the success of inference attacks.
5. The best method of protecting sensitive data is to maintain control of it and not store it in a cloud location. If the data must be shared via a cloud location, the next best protection is to encrypt it by using client encryption methods rather than encryption methods used by the cloud facility.

Objective 6.6: Understand countermeasure principles (e.g., defense in depth)

1. Defense in depth refers to using multiple countermeasures. For example, a firewall and antivirus software on a single system provide more protection than just one or the other. Defense diversity refers to using controls from different vendors for similar controls. For example, if two network firewalls are needed, having network firewalls from two different vendors provides defense diversity. It's unlikely for both firewalls to have exploitable vulnerabilities at the same time.

CHAPTER 7

Operations security

The Operations Security domain focuses on the policies, procedures, and controls used in many of the day-to-day operations of an organization. When preparing for this domain, you should have a good understanding of basic controls specified in written security policies, such as the principle of least privilege, separation of duties, and job rotation. Additionally, you should understand the purpose and methodologies used in several management controls, such as change management and vulnerability management. Detective and preventive controls such as intrusion detection and prevention systems help detect and prevent many attacks, but some still get through. When an incident does occur, an effective incident response plan helps ensure that the damage is minimized and that a repeat occurrence is less likely. This domain also addresses availability concerns through system resilience and fault tolerance requirements.

This chapter covers the following objectives:

- Objective 7.1: Understand security operations concepts
- Objective 7.2: Employ resource protection
- Objective 7.3: Manage incident response
- Objective 7.4: Implement preventative measures against attacks (e.g., malicious code, zero-day exploit, denial of service)
- Objective 7.5: Implement and support patch and vulnerability management
- Objective 7.6: Understand change and configuration management (e.g., versioning, base lining)
- Objective 7.7: Understand system resilience and fault tolerance requirements

Objective 7.1: Understand security operations concepts

Topics in this objective are primarily focused on basic security policies designed to prevent fraud against a company. For most individuals, these policies reduce the temptation to commit fraud by limiting their access and abilities. Additional controls such as least privilege, job rotation, and mandatory vacation policies deter fraud by increasing the risk of being caught. This topic also addresses methods to mark, handle, store, and destroy sensitive information.

Exam need to know...

- **Need-to-know/least privilege**
  For example: What is meant by least privilege? What type of accounts should the principles of need-to-know and least privilege be applied to?
- **Separation of duties and responsibilities**
  For example: What is reduced through an effective separation of duties policy?
- **Monitor special privileges (e.g., operators, administrators)**
  For example: Which accounts should be monitored when monitoring special privileges?
- **Job rotation**
  For example: What is a core security goal related to job rotation?
- **Marking, handling, storing, and destroying of sensitive information**
  For example: What is the primary goal of marking sensitive information? What is the most effective method of sanitizing a drive that includes sensitive information?
- **Record retention**
  For example: What are the risks of not having a data retention policy?

Need-to-know/least privilege

Need-to-know and least privilege are two basic policies designed to ensure that entities have only what is needed to perform tasks and nothing more.

True or false? A least privilege policy is focused on the creation of an account and ensures that the user has only the privileges needed to perform their job.

Answer: *False*. The principle of least privilege should be applied throughout the lifecycle of any account, not just when the account is created.

In general, need-to-know is focused primarily on access to information, and it includes access permissions for files and data within databases. In contrast, the principle of least privilege includes both rights and permissions. Rights refer to an entity's ability to perform actions such as changing configuration settings or installing software.

Further, need-to-know generally refers to an individual and includes permissions granted to an individual's user account. However, the principle of least privilege is applied to all entities, not just user accounts. This includes any service accounts used by applications and services. If the application or service is compromised, the principle of least privilege limits the access and the potential damage from the compromise.

> **EXAM TIP** The principles of least privilege and need-to-know are important concepts that should be included in security policies and tracked throughout the lifecycle of any type of account. When accounts are first provisioned, they should be granted access to only what is needed and no more. Periodic user audits and reviews can discover when individual accounts have been granted excessive privileges after the initial provisioning stage.

> **MORE INFO** Chapter 1, "Access Control," also mentions least privilege, separation of duties, and job rotation concepts in the context of policies. Organizations commonly include these principles in written security policies to ensure that they are implemented throughout the organization.

Separation of duties and responsibilities

A separation of duties and responsibilities policy divides tasking between entities. A primary goal is to ensure that no single entity can control an entire process.

True or false? A separation of duties and responsibilities policy can reduce fraud within an organization.

Answer: *True*. The separation of duties and responsibilities between multiple individuals decreases fraud.

For example, organizations commonly have different people working in accounts receivable and accounts payable. This prevents a single person from submitting a bogus invoice to accounts receivable and then approving its payment from accounts payable.

> **EXAM TIP** A separation of duties policy prevents a single person from having full control over any process and helps reduce the likelihood of fraud. Instead of a single employee being able to commit fraud, two or more employees must collude or conspire to commit fraud when a separation of duties policy is in place. This increases the risk to each employee, helping deter fraud.

Monitor special privileges (e.g., operators, administrators)

Most accounts have basic user privileges, but some accounts are granted significantly more rights and permissions. Because these accounts have additional special privileges, they represent a higher risk to the organization. Specifically, users with these special privileges have access to more data and resources and can cause more damage to the organization.

True or false? Checking system permissions will show who has administrative-level permissions and provide an assessment of whether or not least privilege policies are enforced.

Answer: *True*. By checking administrative and other high-level permissions, you can assess whether or not least privilege policies are being followed.

Individuals with high-level administrative permissions will have full access to systems, but the number of individuals with this high-level access should be limited. Monitoring accounts with special privileges includes two primary tasks:

1. Ensure that membership in these accounts is limited. This is directly related to the principle of least privilege.
2. Monitor usage of these accounts. For example, someone with administrative access can restore files and change permissions. Using these privileges, they can potentially access any data within the organization.

EXAM TIP An access control policy will include direction on how to implement the principle of least privilege, separation of duties, need-to-know, and the granting of special privileges. This is accompanied by audit controls that can track usage of special privileges.

Job rotation

A job rotation policy ensures that users are rotated into different jobs periodically. A primary goal is to reduce risk of fraud that can occur if an individual is the only person that understands a specific job. Job rotation policies also reduce the risk of an organization depending on a single employee for any tasks or job roles.

True or false? Job rotation policies ensure that individuals regularly take vacations.

Answer: *False*. A job rotation policy ensures that individuals are rotated out of jobs periodically with the goal of reducing the risk of fraud. They aren't related to mandatory vacation policies.

Mandatory vacation policies are often used in financial institutions. They require individuals to take two-week vacations, and while individuals are on vacation, someone else must handle their job tasks. This significantly increases the possibility of discovering suspicious activity or fraud while the person is away.

EXAM TIP Job rotation policies help prevent fraud by preventing any single person from controlling the job tasks for too long. Additionally, job rotation policies provide cross-training for employees within an organization and help an organization manage emergency absences for any employee.

MORE INFO The Federal Deposit Insurance Corporation (FDIC) has formally recommended vacations of two consecutive weeks as a security control and recommends that external examiners verify this policy is in place. You can read the FDIC's recommendations here: *http://www.fdic.gov/news/news/financial/1995/fil9552.html*.

Marking, handling, storing, and destroying of sensitive information

A key element of protecting sensitive data is ensuring that it is correctly marked. For example, many companies use clearly visible labels such as public, confidential, secret, or top secret. These labels can be external labels on computers and portable media and can also be digital labels that show up when files are opened.

True or false? A primary goal of marking sensitive information is ensuring that all personnel know the value of the information they are handling.

Answer: *True*. Marking sensitive information with labels makes the type of data being handled clear to all personnel.

> **EXAM TIP** There are several different names used for labels. For example, the United States government uses top secret, secret, confidential, and unclassified labels for data (from most sensitive to least sensitive). Many public organizations use labels such as confidential, private, sensitive, and public (from most sensitive to least sensitive), although there isn't a definitive standard used by all organizations.

Media should always be protected at the same level of the data it contains. For example, magnetic tape backups that contain highly sensitive data should be protected as highly sensitive data. Unmarked media might easily be moved into an unmanned warehouse without adequate protection, but if it is clearly marked, personnel can easily identify the value of the data and are less likely to make such a mistake.

True or false? Magnetic media should be sanitized prior to disposal.

Answer: *True*. Sanitization is the process of destroying all data, including all data remnants, from media prior to disposal.

Simply deleting a file is not an effective method of removing data remnants because there are many programs that can undelete a file. Similarly, formatting a drive isn't effective because unformat programs are available. However, many software programs are designed specifically to overwrite a file or disk. Some will write random patterns of 1s and 0s multiple times whereas others write a specific pattern followed by a complementing pattern. The goal is to destroy all data, including any data remnants.

> **EXAM TIP** It is rarely acceptable to downgrade media to a lower classification, but if this is necessary, the media should first be sanitized using the requirements for the original classification. Simply deleting the original data and putting a new label on the disk is not acceptable because data remnants of the higher-classified data will remain on the media until it is properly sanitized.

True or false? Exposing magnetic tape to a magnetic field is known as degaussing it, and it will sanitize the tape.

Answer: *True*. A degausser is a powerful magnet and will remove all usable data from a tape. You can sanitize a tape with a degausser or destroy the tape to sanitize it. Degaussers can also be used for hard drives, but they destroy the hard drive and make it unusable.

Third-party companies provide disposal services to many organizations. They arrive at the organization with a truck that includes large industrial-sized shredders, and they can destroy the documents on-site. Many shredders can also destroy media such as tapes, CDs, and DVDs. Industrial-sized shredders can destroy hard drives.

True or false? It is possible to purge any type of media by degaussing it.

Answer: *False*. Purging media removes all traces of readable data, and degaussing is an effective purging method for magnetic media but not for other media such as CDs and DVDs. Other purging methods include writing patterns of 1s and 0s over the media to remove the data remnants.

EXAM TIP Degaussing will sanitize magnetic media such as hard drives and tapes. However, it will not sanitize non-magnetic media such as CDs or DVDs. The only guaranteed method of sanitization is destruction, and destruction methods include shredding, disintegrating, incinerating, pulverizing, and melting.

MORE INFO NIST has published SP 800-88, "Guidelines for Media Sanitization," which includes detailed explanations of different types of sanitization, including definitions for disposal, clearing, purging, and destroying. You can access it from the NIST Special Publication page: *http://csrc.nist.gov/publications/PubsSPs.html*.

Record retention

Record retention refers to policies that specify how long data should be retained. In some situations, data retention is governed by law and the retention policy must abide by the law.

True or false? Record retention policies can help limit a company's legal liability.

Answer: *True*. Many organizations implement data and record retention policies to limit their legal liabilities. If the data exists, a court order can order the company to release it. If a policy mandates the deletion of data after a period of time and the data is deleted, the company's liability is limited because the data is gone. If a policy doesn't exist, it can result in a significant amount of labor to review all available data to locate specific data in response to the court order.

It has become very common for organizations to implement specific data retention policies. These apply to all types of data but are often focused on email.

EXAM TIP Record retention policies define how long data is to be kept. They should always comply with governing regulations but should also define the maximum length of time that data should be maintained. This limits data storage requirements and potential legal liabilities.

MORE INFO A white paper titled "Electronic Data Retention Policy" on the SANS Reading Room site provides some good examples of how retained data can cause legal problems. It also includes a list of items to include in a data retention policy. You can access the paper here: *http://www.sans.org/reading_room/whitepapers/backup/electronic-data-retention-policy_514*.

Can you answer these questions?

You can find the answers to these questions at the end of the chapter.

1. What is the purpose of a least privilege policy?
2. How can a separation of duties policy help reduce fraud?
3. Why do audits and reviews focus on accounts with elevated privileges?
4. What is the difference between a job rotation policy and a mandatory vacation policy?

5. What should be done with media prior to disposing of it?
6. Personnel within a company can keep email as long as they want, and many employees have emails from 10 years ago. What should this company implement to reduce their legal liability?

Objective 7.2: Employ resource protection

Resource protection refers to the practices implemented to protect media, software assets, and hardware assets. A significant part of media protection is an understanding of properly marking, handling, storing, and destroying media as covered in the previous objective.

Exam need to know...

- Media management
  For example: What are two primary risks related to USB flash drives?
- Asset management (e.g., equipment lifecycle, software licensing)
  For example: What should be done prior to disposing of equipment?

Media management

Media management refers to the steps taken to protect electronic media an organization maintains. It includes media such as CDs/DVDs, portable hard drives or USB flash drives, and tape media.

True or false? A significant risk when using USB flash drives is data loss.

Answer: *True*. USB flash drives are small enough to fit into a pocket, yet they can store gigabytes worth of data. It's relatively simple for users to copy the data onto the drive and take it out of the organization.

> *EXAM TIP* Written security policies can be combined with technical policies to restrict the use of portable media devices such as USB drives. In addition to protecting against data theft, these policies can also help prevent employees from inadvertently bringing malicious software (malware) into the organization. This can be a significant challenge when an organization supports telecommuting, because employees who work from home often need an acceptable method to transfer data between work and home. One solution is to ensure that all systems have up-to-date antivirus software and that USB flash drives are scanned as soon as they are inserted into a system.

An important part of media management is ensuring that media is properly marked based on the data it contains. When properly marked, it is easier for personnel to recognize the value of the data contained on the media.

> *MORE INFO* When using tapes for backups, media management includes a plan to rotate tapes out of service before they reach the end of their lifecycle. Chapter 8, "Business continuity & disaster recovery planning," discusses some different tape rotation methods, such as the Towers of Hanoi.

Asset management (e.g., equipment lifecycle, software licensing)

Asset management refers to tracking valuable assets throughout their lifetimes. Many organizations use automated inventory systems such as bar code systems and radio frequency identification (RFID) systems to track equipment.

True or false? Checklists should be used to ensure that systems are sanitized prior to disposal.

Answer: *True*. Checklists ensure that systems have been checked prior to disposal. Checklists include items such as verifying that CD/DVD drive bays are empty and that the disk drives do not include sensitive information. Depending on the data and an organization's internal procedures, the disk drives might need to be removed and destroyed, or sanitized with specific software programs to remove data remnants.

True or false? Software keys do not need to be protected as much as the original discs that hold the software.

Answer: *False*. Software keys are more valuable than the discs. In many cases, vendors track usage of a software key and do not allow it to be reused. Software can often be downloaded for free from the Internet but cannot be activated without a key.

For example, an organization could purchase a set of 10 software keys for internal computers. If an employee uses two of these for home computers, these software keys are now considered used and cannot be used on the organization's computers.

Can you answer these questions?

You can find the answers to these questions at the end of the chapter.

1. What is a primary risk when users are able to transfer data files between their work and home computers?
2. Why should disk drives be removed and destroyed from a computer that is being donated to a charity?

Objective 7.3: Manage incident response

A primary goal of security is to prevent security incidents, but they still occur. When an incident occurs, it's important to respond appropriately to limit the potential damage. An effective incident response program includes elements for detection, response, reporting, recovery, and remediation.

Exam need to know...

- Detection
  For example: What is the definition of a security incident? Where should an organization document its definition of a security incident?

- Response
  For example: What is the first goal in response to an incident? Is it appropriate to power off a system as a first response?
- Reporting
  For example: To whom should personnel report an incident? How will personnel determine to whom they report incidents?
- Recovery
  For example: What is involved in the recovery stage if a system is infected with malware?
- Remediation and review (e.g., root cause analysis)
  For example: What is the primary purpose of a review after an incident? If changes were made to a system during recovery, what else should be done?

Detection

In general, an incident is any violation of a security policy. NIST SP 800-61, "Computer Security Incident Handling Guide," defines it as a "violation or imminent threat of violation of computer security policies, acceptable use policies, or standard security policies." Organizations can define it differently, but it's important that it is defined somewhere so that it is clear to all involved when an incident occurs. In many cases, an organization will define an incident in a security policy or in an incident response plan.

True or false? All events are considered security incidents.

Answer: *False.* An event is any observable event, and an adverse event is an event with a negative consequence. However, only adverse events that are computer-security related are considered security incidents.

Natural disasters such as fires and floods are adverse events, and they can impact the availability of systems. However, they aren't considered security incidents.

> **EXAM TIP** Organizations often create an incident response plan to prepare for potential incidents. The plan includes a definition of an incident, how to report an incident, roles and responsibilities for different personnel when responding to an incident, and detailed steps for how to respond to an incident. An incident response team is a group of personnel that will respond to an incident, and they need to be trained on how to respond to different types of incidents.

> **MORE INFO** NIST SP 800-61 is an excellent source for information about incident response. You can access it from the NIST Special Publication page: *http://csrc.nist.gov/publications/nistpubs/*. This document identifies the incident response lifecycle as 1) Preparation, 2) Detection and analysis, 3) Containment, eradication, and recovery, and 4) Post-incident activity.

Response

After an event has been confirmed as an incident, the next step is to respond to the incident. A primary goal of the response is to mitigate the scope of the incident.

True or false? After verifying that a server is infected with a worm, the server should be isolated.

Answer: *True*. Containing the scope of a security incident is extremely important, and a server can be isolated by blocking all outbound connections, disabling the network interface card (NIC), or simply removing the network cable.

A worm will automatically try to connect with other systems on the network and infect them. Isolating systems that are infected with a worm prevents the infection from spreading.

> **EXAM TIP** Containment of an incident should be a priority as soon as an event has been verified to be a security incident. Isolating a system from a network is a common way to isolate an incident, but in some situations, it might require the isolation of an entire network. This step should also consider the potential need to gather evidence from the system. If a forensic analysis team needs to gather evidence, the system should not be shut down or manipulated. Removing the network cable will isolate the system without affecting any potential evidence.

> **MORE INFO** Chapter 9, "Legal, regulations, investigations, and compliance," covers forensic procedures that are important to understand when responding to an incident. Evidence is contained in volatile memory and can be retrieved as long as the computer is not turned off. Opening and accessing files that an attacker accessed also modifies evidence because the system will have a record of when the last user accessed the file but not necessarily when or if the attacker accessed it. In general, if evidence needs to be collected, nothing should be manipulated until a forensic specialist has collected the evidence.

Reporting

Reporting includes notification of several different people, depending on the severity of the incident. First, personnel on the incident response team need to be notified. If the incident is severe, management and executive personnel need to be notified. In some cases, customers and/or legal authorities will also need to be notified.

True or false? As long as the incident isn't public, it's not necessary to report any breaches associated with customer information.

Answer: *False*. Personally identifiable information (PII) is protected by many different laws. At the very least, customers need to be notified when PII about them has been breached.

An organization risks the loss of customer goodwill when customer PII is compromised. Organizations aren't necessarily anxious to release this information, but when they are proactive, they often show they have better control of the situation, even after a data breach. This can help to minimize long-term negative publicity.

> **EXAM TIP** An incident response plan should outline who should be informed or notified based on the severity of an incident. An organization has a legal responsibility to report the breach of any losses related to PII, but the organization should ensure that this is done in such a way that it minimizes loss of customer confidence.

MORE INFO NIST SP 800-122, "Guide to Protecting the Confidentiality of Personally Identifiable Information (PII)," provides detailed information about PII and includes a section on incident response for breaches involving PII. You can access it from the NIST Special Publication page: *http://csrc.nist.gov/publications/nistpubs/*.

Recovery

Recovery steps remove any damage caused by the incident and return it to full service. Sometimes this can be as simple as restarting the system. In other cases, it might require completely rebuilding the system from scratch.

True or false? When an incident involves malware, recovery includes eradication of the malware.

Answer: *True*. Recovery includes eradication of all elements of the incident, including when the incident involves malware.

> ***EXAM TIP*** Eradication of malware might be the removal of malicious code from infected files, deleting infected files, and/or restoring files with uninfected versions. The specific procedure is dependent on the type of malware and the damage it caused.

When multiple systems are affected, recovery is often performed in stages. For example, if multiple systems are infected with a worm, all infected systems must first be isolated. Then, based on the priority of the systems, individual systems are recovered by eradicating the malware on each. One by one, the recovered systems are added back to the network. If systems are added back to the network before the malware has been removed, they can easily infect all systems on the network again.

Remediation and review (e.g., root cause analysis)

The last stage of an incident response examines the incident with the goal of identifying what happened and why it succeeded. Ideally, a review can identify the root cause of the incident. When the cause is understood, it's much easier to implement controls to prevent a repeat occurrence.

True or false? A post-incident review should include a review of actions taken by the team, with the primary goal of assigning blame.

Answer: *False*. A "lessons learned" review is an important step in a review, but the goal is not to assign blame. Instead, the review should identify what worked and what didn't, with the goal of improving the incident response plan.

> ***EXAM TIP*** Reviewing the overall incident and response is an invaluable step in improving the incident response plan. It helps the organization understand exactly what happened and how well the organization was able to respond. In some cases, personnel might follow the plan but find that it wasn't effective. In other cases, personnel might not have the proper knowledge or training to properly implement the steps in the plan. By reviewing the incident, the organization can resolve any problems that allowed the incident to occur and can improve the team's response to any subsequent incidents.

True or false? Emergency changes made during an incident should be documented and reviewed within a change management process.

Answer: *True*. Emergency changes often bypass a change management process, but they should be reviewed after the incident to ensure that an emergency fix doesn't cause other problems.

> **EXAM TIP** Emergency changes might be appropriate to keep systems protected from an incident. However, these changes should be included in any change management and configuration management documentation. If a system fails and needs to be rebuilt for some other reason, up-to-date configuration management documentation ensures that the system is configured in a secure manner.

Can you answer these questions?

You can find the answers to these questions at the end of the chapter.

1. What is the difference between an adverse event and a security incident?
2. What is the easiest way to contain a virus on a networked server?
3. A security breach has resulted in the compromise of customer accounts, including their user names and passwords. Who must be informed?
4. What is a likely outcome of a root cause analysis?

Objective 7.4: Implement preventative measures against attacks (e.g., malicious code, zero-day exploit, denial of service)

This objective requires a basic understanding of many different types of attacks but focuses on the methods to prevent them. Preventative measures help to prevent, detect, and block attacks. Two primary methods of preventing successful attacks are the use of antivirus software and the use of intrusion detection and prevention systems.

Exam need to know...

- Attacks
  For example: What's the difference between spyware and a virus? What is an anonymous relay?
- Preventative measures
  For example: What is the major disadvantage with an intrusion detection system (IDS). What must be created to support a behavior-based IDS?

Attacks

Attacks come in many different forms and have many different goals. Denial of service (DoS) and distributed denial of service (DDoS) attacks attempt to disrupt a server or service from providing a service. Many malware attacks infect a system and grant the attacker remote control to a system. The attacker can monitor activity to steal data and credentials and/or join the system to a botnet.

True or false? A zero-day exploit takes advantage of a previously unknown vulnerability.

Answer: *True*. Zero-day exploits or attacks take advantage of vulnerabilities that are not widely known.

There isn't a definitive definition of a zero-day exploit, but in general, it refers to an attack on a vulnerability that is not known. This includes vulnerabilities not known by the original developer or the public. Imagine that an attacker learns about a vulnerability on January 1 and begins exploiting it. Then, on January 15, the developer learns about the vulnerability and begins developing an update to correct it. Attacks during these 15 days (January 1 to January 15) are considered zero-day exploits.

Now imagine that the developer creates and releases an update on February 1. The general public is now aware of the vulnerability and the update, but they weren't aware of it previously. Many people consider attacks between January 16 and February 1 to also be zero-day exploits because the vulnerability was not previously known by the public. In contrast, any attacks against this vulnerability after February 1 are not zero-day exploits. An update is available that can block the attack, and as long as systems are patched, they are protected.

> **EXAM TIP** Many general security practices help reduce vulnerabilities to zero-day exploits. This includes removing or disabling all unneeded services and protocols, enabling firewalls on all systems, and using up-to-date antivirus software. Behavior-based intrusion detection and prevention systems combined with honey pots or honey nets are also useful in identifying suspicious activity related to zero-day exploits.

> **MORE INFO** SearchSecurity includes a definition of a zero-day exploit on its website (*http://searchsecurity.techtarget.com/definition/zero-day-exploit*) as an attack that occurs on the same day that the vulnerability is known. Webopedia has a slightly different definition: *http://www.webopedia.com/TERM/Z/Zero_Day_exploit.html*. Wikipedia has a more comprehensive definition: *http://en.wikipedia.org/wiki/Zero-day_attack*.

True or false? Spyware allows an attacker to capture personal information from a computer.

Answer: *True*. Many different types of spyware are available, and they enable the attacker to capture information from the user's computer.

> **EXAM TIP** An understanding of different types of malware is essential when studying for the CISSP exam. Malware includes viruses, worms, logic bombs, Trojan horses, and spyware. Viruses must be executed through some type of user interaction, while worms can infect other systems on a network without user interaction. Logic bombs run in response to some event, such as a time or date. A Trojan horse looks like something appealing to the user but has a hidden malicious component. Spyware is installed without the user's knowledge and can gather personal information, including browsing habits and user name/password combinations.

True or false? Automated methods that identify NICs that are operating in promiscuous mode can help prevent sniffing attacks.

Answer: *True*. Sniffers (also called protocol analyzers) must operate in promiscuous mode to capture all traffic received at the NIC (not just traffic sent to or from the NIC), and it is possible to check NICs to determine whether they are operating in promiscuous mode.

> **EXAM TIP** Unauthorized sniffers can capture data sent across a network. If the data is sent in an unencrypted format, the attacker can read the data as clear text. An additional protection is to encrypt all sensitive data transmitted on a network.

True or false? Anonymous relays should be disabled on email servers.

Answer: *True*. Email servers will relay email from other systems, and if anonymous relay is enabled, a spammer can use the email server to relay spam. The spam then appears to be coming from the server with anonymous relay enabled.

> **MORE INFO** Chapters 1 and 2 cover many different types of attacks related to networks, such as DoS, DDoS, and spoofing attacks. Many network attacks include spoofing techniques by using a forged source IP address.

Preventative measures

Preventative measures are designed to prevent attacks from succeeding or, at the very least, notifying personnel when a suspected attack is in progress. IDSs, intrusion prevention systems (IPSs), firewalls, and antivirus software are core preventative measures.

IDSs monitor networks and systems for malicious activity, including potential attacks, breaches, and other unauthorized access to systems. They are configured to alert on specific activities based on preset thresholds. For example, they can detect when a potential SYN flood attack is occurring, based on the number of incomplete TCP sessions on a server. When the threshold is reached, IDSs log the event and raise an alarm by sending a notification.

> **EXAM TIP** One of the major disadvantages with IDSs is the potential number of false positives. When the threshold is set too low, it continuously alerts on legitimate activity. Each alert should be investigated, and each false positive takes administrators away from other work. With too many false positives, administrators stop paying attention to the IDS and actual attacks are ignored.

True or false? A signature-based IDS is used to detect known attack methods.

Answer: *True*. Signature-based detection uses a database similar to how antivirus software uses a database of signatures.

IDSs use signature-based detection or anomaly-based detection. Anomaly-based detection uses statistical analysis. It starts by creating a baseline of normal activity and then monitors activity against the baseline. When the traffic deviates significantly from the baseline, it sends an alert. If the network is updated, the baseline must also be updated.

> **EXAM TIP** The two methods of detection used by IDSs and IPSs are signature-based and anomaly-based. Signature-based methods detect known attack methods and must be updated regularly. Anomaly-based detection methods detect attacks by noting significant changes in network behavior from a baseline. After network changes are implemented, a new baseline needs to be created.

> **MORE INFO** Chapter 4, "Software development security," discusses buffer overflow attacks, which attempt to take advantage of poor coding practices by sending unexpected input to a program. Input validation techniques are an important preventative measure to prevent buffer overflow attacks. Additionally, intrusion detection systems can detect buffer overflow attack attempts based on attack signatures and can take steps to block them.

True or false? An IDS can detect attacks but cannot detect the source of the attack.

Answer: *True*. An IDS can detect attacks, and it can log the source IP address. However, it cannot determine whether the source IP address is the actual address or a spoofed IP address.

> **MORE INFO** IDSs and IPSs are mentioned in the Chapter 1 in the context of logging and monitoring. A primary difference between an IDS and an IPS is that an IPS is placed inline with the traffic. That is, all traffic must go through an IPS. The IPS can filter out malicious traffic, preventing it from reaching the internal network.

An IDS is identified as either passive or active, based on how it responds.

- A passive response provides notification to personnel. This might come in the form of an email, text message, or some other type of alert. Personnel can then investigate the activity and determine whether action is needed.
- An active response provides notification to personnel and also takes action to block the attack. For example, it might change the configuration of a router or a firewall to block all traffic from the source IP address, or it might block specific types of protocol traffic.

True or false? Antivirus software should be installed at the network boundary and on all internal systems.

Answer: *True*. Antivirus software at the network boundary can filter and remove malware before it enters the network. Installing antivirus software on all systems ensures that they are protected if malware enters the network from a different path, such as from a USB flash drive.

Many server applications require specialized antivirus software. For example, email and database servers require specialized software to detect and remove malware. Regular antivirus software can easily corrupt the databases used by these servers.

EXAM TIP The primary protection against malware is up-to-date antivirus software. Most current antivirus software has the ability to regularly check for new signatures and automatically download them. Corporate versions allow administrators to control when signatures are downloaded and when they are pushed to clients.

Can you answer these questions?

You can find the answers to these questions at the end of the chapter.

1. How long can an attacker use a zero-day exploit?
2. What's the primary difference between a virus and a worm?
3. Administrators have modified some network hardware. What must they update to ensure that a behavior-based IDS can accurately detect anomalies?

Objective 7.5: Implement and support patch and vulnerability management

Patch management ensures that systems are kept up to date with current updates, and vulnerability management regularly tests systems for potential vulnerabilities. Many patch management systems include the ability to check systems to ensure that they are up to date. Vulnerability assessment software can also detect unpatched systems.

Exam need to know...

- Patch management
  For example: What should be done before deploying an update? How should updates be approved?
- Vulnerability management
  For example: What is the primary output of a vulnerability assessment? What is the difference between a vulnerability assessment and a penetration test?

Patch management

Throughout the lifecycle of any operating system or application, various problems are discovered. Some of these are minor issues that affect a software feature on a limited number of systems, and some are major issues that can affect the security of a large number of systems. Vendors regularly create and release updates, and patch management practices ensure that systems are kept up to date.

True or false? All updates should be applied as soon as possible after they are released.

Answer: *False.* Updates should first be evaluated to determine whether they are needed, because all updates are not necessarily needed for all systems.

True or false? Updates should be tested on nonproduction systems prior to deploying them.

Answer: *True*. Testing updates on nonproduction systems helps identify potential problems without impacting productivity.

In some cases, updates have had unexpected results due to conflicts with other software or hardware. In the worst case scenarios, updates have prevented the systems from operating until the update was removed. Testing updates on nonproduction systems prevents the updates from impacting the availability of production systems.

Updates with the potential for the most serious security risks are placed higher in the priority list within the testing process. Some small organizations choose to automatically install updates as they become available. This reduces administrative overhead, but the risk is that an update might cause problems for all the systems at the same time.

Larger organizations use automated processes to deploy the updates. They can use these processes to automate the deployment to test systems first, and then, after testing, they can automate the deployment to production systems. Automated systems include the ability to regularly audit systems to verify that they are patched.

> **MORE INFO** There are several tools available to automate the deployment of patches. Microsoft tools include Windows Software Update Services (WSUS), which is free and used in many small-to-medium-sized organizations. System Center Configuration Manager (SCCM) is a more advanced tool that can schedule the deployment of updates. SCCM can also deploy images and applications to systems. Third-party tools are also available for both Microsoft and non-Microsoft environments.

True or false? Patch management changes should be submitted through a change management process.

Answer: *True*. Updates are designed to correct a problem in the code, but the updated code can also cause undesired results. Submitting these changes through the change management process ensures that they are examined, tested, and documented.

> **EXAM TIP** A core method of keeping systems secure is keeping them up to date. When vulnerabilities are known, vendors develop and release patches, but systems remain unprotected until they are patched. Attackers commonly look for unpatched systems and exploit them.

Vulnerability management

Vulnerability management is the practice of regularly scanning systems for known vulnerabilities. A vulnerability assessment tool will scan the network and systems but does not attempt to exploit any vulnerabilities.

True or false? The primary difference between a penetration test and a vulnerability assessment is that a penetration attempts to exploit vulnerabilities.

Answer: *True*. Vulnerability assessment tools do not attempt to exploit vulnerabilities. A penetration test often uses vulnerability assessment tools to discover vulnerabilities and then attempts to exploit the vulnerabilities.

The outcome of a vulnerability assessment is a report provided to management personnel. Management decides which vulnerabilities to mitigate and which vulnerabilities to accept.

> **EXAM TIP** Security personnel often run vulnerability assessment tools on a regular basis to detect vulnerabilities. They should never attempt penetration tests without express approval from management. Unauthorized penetration tests can impact the performance of a network or system and be viewed as a malicious attack. In many organizations, employees are fired if they are caught performing internal penetration tests without authorization.

> **MORE INFO** The MBSA tool is available as a free download here: *http://www.microsoft.com/download/details.aspx?id=7558*. It can scan multiple systems, check for vulnerabilities, and provide a detailed report on known vulnerabilities. Nessus is a third-party tool that has the capabilities to scan multiple systems and check for vulnerabilities. You can get an evaluation copy here: *http://www.tenable.com/products/nessus*.

True or false? A penetration test where the testers do not have any knowledge of the network is also known as a black box test.

Answer: *True*. When penetration testers do not have any knowledge about a target, it is known as a black box test. In some documentation, this is also referred to as a zero knowledge test.

> **EXAM TIP** Penetration testers can have full knowledge of a target for white box testing, partial knowledge of a target for gray box testing, or zero knowledge of a target for black box testing.

Can you answer these questions?

You can find the answers to these questions at the end of the chapter.

1. What are different elements of a patch management process?
2. Are all vulnerabilities reported from a vulnerability assessment mitigated?
3. What information is available to a tester in a block box test?

Objective 7.6: Understand change and configuration management (e.g., versioning, base lining)

Change and configuration management work together to ensure that systems are configured correctly when deployed and remain that way through their lifetime. Many disastrous IT outages have been traced back to unauthorized changes, often performed by well-meaning technicians and administrators. However, systems are complex and a simple change in one area can result in significant problems in another area. Change management processes ensure that these changes are examined, approved, and documented before they are implemented. Configuration management processes focus on the deployment of systems to ensure that they are deployed in a consistent, secure manner. During the lifetime of a system, change management processes are integrated with the configuration management documentation to ensure that approved modifications are documented.

Exam need to know...

- Change management
  For example: What types of changes should be submitted to a change management process prior to making the change?
- Configuration management
  For example: What is a method of ensuring that all systems start with a similar baseline configuration?

Change management

Change management is a group of practices that ensure that all changes are evaluated, documented, and tracked. One of the primary goals is to prevent unintended outages caused by unauthorized changes.

True or false? Change management policies ensure that changes and change requests are documented.

Answer: *True*. Change management policies and procedures ensure that all change requests follow a controlled process. This process includes documenting the change from the original request through the approval and implementation.

> **EXAM TIP** Unauthorized changes often cause outages that are preventable with change management practices. All changes should be submitted through a change management process, from simple configuration changes to major system changes, because any change has the potential to cause an unintended outage. Change management practices help prevent these unintended outages by ensuring that all changes are first requested and examined before they are approved and implemented. They also provide a streamlined process for all change requests. The only exceptions are emergency changes completed during a security incident, but these changes should still be submitted through change management when the incident has been resolved.

True or false? A change advisory board (CAB) is responsible for assessing, prioritizing, and scheduling changes.

Answer: *True.* CAB is specific terminology identified in Information Technology Infrastructure Library (ITIL) documentation. Many organizations use change management practices but don't necessarily implement formal ITIL procedures. These organizations still designate a group of people who perform similar functions as the CAB but might be called something else.

> **MORE INFO** Change management practices are often associated with the ITIL. The ITIL books are expensive and not easily accessible, but there are other resources that explain similar concepts. For example, Microsoft has adopted the Microsoft Operations Framework (MOF), which uses four phases of the IT service lifecycle. The Manage layer includes both change and configuration management. You can view the entire MOF library at *http://technet.microsoft.com/library/ee923730.aspx*, and you can view topics on the Manage layer at *http://technet.microsoft.com/library/cc506048.aspx*.

True or false? Changes should be tested on all production equipment before being implemented.

Answer: *False.* Whenever possible, changes should be tested on nonproduction equipment prior to making a change. Testing a change on a production system might result in an unintended outage, and the goal of change management is to avoid unintended outages.

> **EXAM TIP** Testing changes is an important part of change management. The overall process consists of a formal request for the change, a review of the change, testing the change, approving the change, implementing the change, and reporting when the change is completed. Each of these steps is documented within the change management system, but different organizations use a variety of different methods to document changes. However, any change management system should be able to track different versions of changes to a system and the status of each change at any given time. Additionally, this documentation should be backed up and accessible so that it can be accessed during the recovery phase of disaster recovery.

Configuration management

Configuration management practices ensure that systems are deployed in a consistent, secure state and stay that way through their lifetimes. In some organizations, change and configuration management are treated as a single program and named something like change/configuration management.

True or false? System images are a common method of ensuring that multiple systems are deployed in a similar, consistent state.

Answer: *True.* After building a reference computer with an operating system, applications, and security settings, administrators capture an image of the system. This image can then be deployed to multiple computers, and each computer has the same image.

Imaging ensures that important computer and security settings are configured identically on each system. This is also referred to as creating a baseline or starting point for each system.

> **EXAM TIP** Configuration management ensures that systems are deployed in a consistent, secure state. It is integrated with change management practices to ensure that all approved changes are documented as legitimate configuration changes. If a system needs to be rebuilt, the configuration documentation includes the original settings and all approved changes.

It's also possible to use automated methods to check the configuration of systems after they've been deployed. For example, Group Policy in Microsoft networks is used to ensure that systems are configured consistently. An administrator can set the Group Policy setting one time, and the setting applies to all the computers in the scope of the policy. Group Policy settings can be configured to apply to all the computers in the organization or only to specific computers, depending on the needs of administrators. This is an effective method for configuring both desktop computers and servers. Additionally, Group Policy periodically checks the settings on the systems, and if settings are modified, Group Policy resets them.

> **EXAM TIP** Baselining is the process of ensuring that systems start with a common baseline, or common starting point. Both the use of images and Group Policy are baselining methods. While it's possible to have technicians manually configure settings, it isn't as reliable as automated methods.

Can you answer these questions?

You can find the answers to these questions at the end of the chapter.

1. What is a common adverse result when an organization doesn't follow a change management process?
2. What types of changes should be documented?
3. What phase of change management is intended to discover potential problems?
4. What is a benefit of imaging?

Objective 7.7: Understand system resilience and fault tolerance requirements

System resilience and fault tolerance refers to a system's ability to experience a fault but continue to operate. There are multiple methods available to provide system resilience and fault tolerance. From a computer or server perspective, the primary fault tolerant method is to use redundant array of independent disks (RAIDs) to ensure that the disks continue to operate even if a disk fails. When protecting services provided by a server, failover clusters are used to ensure that the service continues to operate even if a server fails.

Exam need to know...

- Fault tolerance for disks
  For example: What is included in a RAID-1 with duplexing? What is the minimum number of disks in a RAID-10?
- Fault tolerance for servers
  For example: What is the primary goal of a failover cluster?

Fault tolerance for disks

Fault tolerance for disks is implemented with different RAID configurations. RAID-0 does not provide any fault tolerance, but it does provide improvements in read and write speed for a disk.

True or false? A RAID-1 always has two disks.

Answer: *True*. A RAID-1 is also known as a mirror and always has two disks.

Figure 7-1 compares a single disk without RAID with a RAID-0 and a RAID-1.

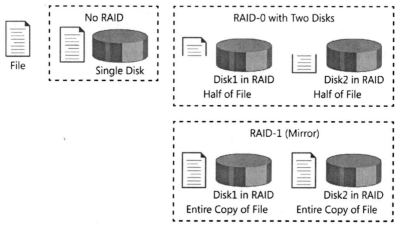

FIGURE 7-1 A single disk without RAID compared with RAID-0 and RAID-1.

RAID-0 (also called striping) includes at least two disks but does not provide fault tolerance. Data is striped across the disks, which improves read and write performance.

RAID-1 (also called mirroring) includes two disks. Anything written to one disk is also written to the other disk. If one of the disks fails, the other disk has all the data.

True or false? RAID-1 with duplexing uses two hard drives and two disk controllers.

Answer: *True*. RAID-1 with duplexing is the same as RAID-1, except that it includes an extra disk controller for each disk. It eliminates the disk controller as a single point of failure.

True or false? A RAID-5 uses at least three disks and can continue to operate if one of the drives fails.

Answer: *True*. RAID-5 uses at least three disks, and the equivalent of one of the disks is devoted to parity information. If one of the drives fails, the RAID can continue to operate.

Figure 7-2 compares RAID-3 and RAID-5. Each of these configurations can survive the failure of a single disk. However, if two or more disks fail, the data is no longer available.

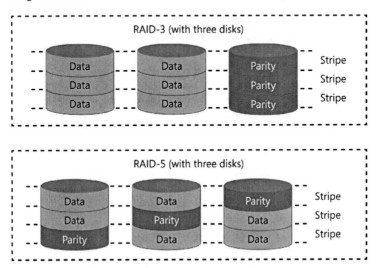

FIGURE 7-2 Comparing RAID-3 and RAID-5.

RAID-3 (also called striping) includes at least three disks, and the last disk is dedicated to parity. Data is written to the first two drives, and the parity data is written to the last drive. If any single drive fails, the system can recalculate the value of the data on the missing drive and continue to operate.

RAID-5 (also called striping with parity) includes at least three disks, and the equivalent of one disk is dedicated to parity. Data is striped across all the drives in 64-KB blocks, and each stripe includes parity data stored on one of the drives. RAID-5 provides better performance gains than does RAID-3, but you might see RAID-3 systems mentioned occasionally.

True or false? A RAID-10 uses at least four disks and, in some cases, can continue to operate if two drives fail.

Answer: *True*. RAID-10 uses at least four disks configured as a stripe of mirrors and can continue to operate if more than one disk fails.

Figure 7-3 shows a RAID-10 with six disks. It's also possible to have a RAID-10 with four disks, eight disks, or any multiple of two disks as long as there are four or more.

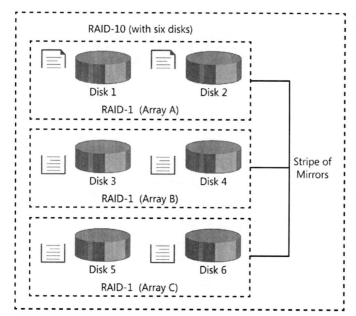

FIGURE 7-3 RAID-10.

RAID-10 (also called a stripe of mirrors) includes several pairs of disks, with each pair configured as a RAID-1 or mirror. If either disk in each pair fails, the other disk has the data and the RAID-10 can continue to operate.

A RAID-10 (also called RAID 1 + 0) can have failures in multiple drives as long as both the drives in any individual mirror do not fail at the same time. For example, if Disk 1, Disk 3, and Disk 5 fail, the RAID can continue to operate because Disk 2, Disk 4, and Disk 6 all include the missing data. However, if Disk 1 and Disk 2 both fail, the entire RAID will fail, because all the data from Array A is no longer available.

> **EXAM TIP** RAID 1, 5, and 10 are common RAID configurations that provide fault tolerance. Make sure you understand how many disks each can use when preparing for the exam. Only RAID-1 uses additional disk controllers to provide disk duplexing. RAID-10 provides the best performance and fault tolerance gains but is also the most expensive.

> **MORE INFO** If RAID configurations aren't clear to you, read this easy-to-understand tutorial on different RAID levels: http://www.acnc.com/raid.

Fault tolerance for servers

The primary method of providing fault tolerance for services is with failover clusters. A failover cluster ensures that a service can continue to operate even if a server fails.

True or false? It's common to use fault tolerant solutions in all servers.

Answer: *False*. Fault tolerant solutions are usually reserved for the mission-critical systems. The added expense of fault tolerant solutions is not justified in many systems used in day-to-day computing.

A simple failover cluster can include two servers. One server is online, and the other server monitors the online server. If the online server fails, the second server senses the failure and automatically takes over. From a user's perspective, there is a momentary delay but no loss of service. More complex failover clusters can include multiple servers. Each server in the cluster should have the ability to take over if any of the other servers fail.

> **EXAM TIP** Failover clusters are expensive. Each server in the cluster needs to have almost identical hardware, and often the equivalent of one server is not being used to provide any services other than monitoring the online server. The benefit is high availability, but the tradeoff is a higher cost.

> **MORE INFO** Microsoft server products support failover clusters in a wide variety of different configurations. The Failover Clustering Overview page provides an overview of failover clustering and how it is used with Windows Server 2012, along with links to many other failover clustering pages. You can access it here: *http://technet.microsoft.com/library/hh831579.aspx*.

True or false? A primary goal of fault tolerance methods is to increase availability.

Answer: *True*. Fault tolerance methods ensure that systems will continue to be available after a fault occurs.

> **MORE INFO** System reliance and fault tolerance methods can also apply to power and communication lines. Chapter 10, "Physical (environmental) security," discusses uninterruptible power supplies (UPS) designed to provide power during short-term power interruptions, and generators designed to provide long-term power. Additionally, organizations can choose to add redundant communications lines to ensure that they have Internet connectivity or connectivity to other locations, even if one of the communications lines fails.

Can you answer these questions?

You can find the answers to these questions at the end of the chapter.

1. What can you add to a RAID-1 to increase availability?
2. How many disks can fail in a RAID-5 before data is no longer available?
3. What should be added to ensure a service will continue to operate even if a server fails?

Answers

This section contains the answers to the "Can you answer these questions?" sections in this chapter.

Objective 7.1: Understand security operations concepts

1. A least privilege policy helps ensure that entities do not have excessive privileges. It does so by mandating that accounts, personnel, software, and services have only the privileges they need to perform their jobs and no more.
2. A separation of duties policy helps reduce the risk of fraud by ensuring that a single person cannot commit fraud alone. It requires two or more people to conspire or collude against the organization. This collusion increases the risk to employees of getting caught and serves as a deterrent to employees who might otherwise consider defrauding the organization.
3. Accounts with elevated privileges can cause the most damage if their owners abuse their privileges. Monitoring all accounts generates a lot of auditable events, but when the focus is on privileged accounts, it's easier to complete a meaningful audit.
4. A job rotation policy rotates people into different jobs on a semi-permanent basis. A mandatory vacation policy requires employees to regularly take a vacation so that someone else fulfills their job duties on a temporary basis. Both can help deter and detect fraud.
5. Media should be sanitized prior to disposal. Organizations will have different policies based on the sensitivity of the data. The most effective method of sanitization is destruction.
6. A data retention policy will limit how long data can be kept and can reduce an organization's legal liability. Even if an organization has done nothing wrong, it can be ordered to maintain data indefinitely in response to a frivolous lawsuit. Similarly, it can be ordered to sift through millions of files looking for specific data in response to a court order. However, with a data retention policy, the effect of these scenarios is limited.

Objective 7.2: Employ resource protection

1. A primary risk of allowing users to transfer data between home and work computers is that they can inadvertently transfer malware.
2. If the disk drives include any type of sensitive, proprietary, or private information, the disk drives should be sanitized. For highly sensitive information, an organization can dictate that the disk drive must be destroyed to sanitize it.

Objective 7.3: Manage incident response

1. An adverse event includes natural events such as fires and floods. A security event does not include natural events but instead is focused on computer-related security events, such as attacks.

2. The simplest way to isolate a networked system is by removing the network cable. This contains the incident on the system but doesn't risk modifying any evidence.
3. This is a compromise of customer PII, and the customer must be informed. Depending on company policies and the organization's jurisdiction, other personnel within the company and external to the company will also be informed.
4. A likely outcome of a root cause analysis is the discovery of a vulnerability that allowed an incident to occur and of possible weaknesses in the response plan. Based on what is learned, controls might be added to eliminate the vulnerability, the plan might be updated to improve the response, or personnel might be provided with additional training.

Objective 7.4: Implement preventative measures against attacks (e.g., malicious code, zero-day exploit, denial of service)

1. Attackers can use zero-day exploits as long as the vulnerability remains. After the developer learns of the exploit, develops an update, and releases it, the vulnerability is no longer considered a zero-day exploit. The time frame between when the attacker learns of the exploit and the developer releases a patch might be days, months, or even years. After an update is released, customers who don't install the update are still vulnerable to the original exploit even though it is no longer considered to be a zero-day exploit.
2. A virus must be executed through some type of user interaction. A worm does not require any user interaction but instead can travel over a network and infect other systems.
3. The baseline for a behavior-based IDS must be updated if the network is modified. A behavior-based IDS identifies potential attacks by comparing a baseline of normal behavior with current activity, but if the network is modified, the baseline is no longer accurate.

Objective 7.5: Implement and support patch and vulnerability management

1. A patch management process will evaluate updates as they become available, submit relevant updates through a change management process, test the updates, deploy the updates, and then audit systems to ensure that they remain updated.
2. A vulnerability assessment detects vulnerabilities, but management within an organization must decide which vulnerabilities to mitigate. It is not possible to eliminate all vulnerabilities, just as it is not possible to eliminate all risks.
3. In a black box test, testers have zero inside knowledge about the target. All the knowledge they have is publicly available when they start. A common practice during a black box penetration test is to use social engineering techniques to get more information about the target.

Objective 7.6: Understand change and configuration management (e.g., versioning, base lining)

1. A common adverse result when an organization doesn't use change management processes is the occurrence of unintended outages. To help avoid unintended outages, change management ensures that changes are examined and tested before implementation.
2. All IT changes should be part of the change management process, and all changes should be documented.
3. The testing phase of change management is intended to discover problems. Whenever possible, changes should be tested on nonproduction systems first.
4. Imaging techniques allow administrators to deploy multiple systems by using identical configuration and security settings. This reduces the workload while also increasing the security on the deployed systems.

Objective 7.7: Understand system resilience and fault tolerance requirements

1. You can add disk controllers to a RAID-1 to eliminate the disk controller as a single point of failure. This is also known as RAID-1 with duplexing.
2. A RAID-5 requires at least three disks, and it can survive the failure of a single drive. If two or more disks fail, the RAID-5 fails.
3. Adding a failover cluster to a network will allow a service to continue to operate even if a single server fails. When a server fails, the service fails over to a different server in the failover cluster.

CHAPTER 8

Business continuity & disaster recovery planning

Business continuity refers to the ability of an organization to continue to operate after a disruption in service. Typically, business continuity addresses significant disasters such as fires and floods, but it also addresses any single point of failure that can prevent a critical business function from continuing. Effective business continuity planning includes several phases and starts by identifying the scope of the plan. A business impact analysis identifies critical business functions, maximum tolerable downtimes, and recovery objectives. Recovery strategies identify the details necessary to recover a critical business function, returning it to service within the time frame identified by the recovery objectives. Plans are tested by using multiple methods and are regularly reviewed to ensure that they still meet the needs of the organization.

This chapter covers the following objectives:

- Objective 8.1: Understand business continuity requirements
- Objective 8.2: Conduct business impact analysis
- Objective 8.3: Develop a recovery strategy
- Objective 8.4: Understand disaster recovery process
- Objective 8.5: Exercise, assess and maintain the plan (e.g., version control, distribution)

Objective 8.1: Understand business continuity requirements

Business continuity planning starts with a project plan to identify the overall scope of the project and milestones within the plan. A business continuity plan (BCP) manager or BCP coordinator develops the initial plan with support from executive management and is responsible for all phases of the plan.

Exam need to know...

- Develop and document project scope and plan
  For example: What does the scope define? What are the overall phases of a BCP project plan?

Develop and document project scope and plan

Business continuity and disaster recovery is encompassed by a BCP, but it takes a lot of pre-planning to create the BCP. One of the first steps is to identify the scope of the plan and create a project plan identifying specific tasks and timelines.

True or false? A BCP focuses only on the business and not on the people.

Answer: *False*. Because BCPs are commonly addressing a response to disasters, they must also include elements to address the safety of personnel. For example, an approaching hurricane might require the company to assist with evacuation. Similarly, if the organization houses any hazardous chemicals or supplies, the scope of the BCP should address steps to prevent such materials from causing damage to personnel.

The creation of the BCP is often divided into five separate phases, and each of these phases would be identified in the project plan. The phases are as follows:

1. **Identify the project scope and create a project plan.** A BCP might be designed to cover the entire organization or only a specific department or location, depending on the size of the organization. Similarly, it might be designed to cover specific disasters or any event that can affect critical business functions. The project plan provides a timeline with specific milestones. Based on what is discovered during the BIA, these milestones might be updated.

   EXAM TIP Personnel involved in the development of any BCP or disaster recovery plan (DRP) must be knowledgeable about the organization's business processes. Without an understanding of how the business operates, it is not possible to accurately identify the critical business factors.

2. **Complete a BIA** A primary purpose of the BIA is to identify and prioritize critical business functions and define maximum tolerable downtimes (MTDs). This information is used to define specific recovery objectives for the critical business functions. For example, an e-commerce website might have an MTD of one hour, requiring a recovery time objective (RTO) of one hour or less.

3. **Develop recovery strategies** Recovery strategies are created to meet the RTOs identified from the BIA. These range from simple backup and recovery procedures for some business functions to the implementation of a hot site as an alternate location for other business functions.

4. **Create the BCP and DRPs** A BCP typically includes multiple disaster recovery plans (DRPs). For example, one DRP might involve the procedures necessary to restore a single critical server, and another DRP might include the overall steps an organization takes to activate an alternate site in response to an approaching hurricane or after an earthquake.

5. **Test and maintain the BCP**   Multiple types of tests validate the plans and ensure that they will work as intended. Periodic reviews verify that the plans cover all of the critical business functions currently owned by the organization and that the recovery strategies still meet the recovery requirements.

MORE INFO   There isn't a definitive document that defines how many steps or stages are required for business continuity planning. The previous five stages map loosely to the objectives in the Business Continuity & Disaster Recovery Planning domain, but other sources list the stages differently. For example, NIST SP 800-34, titled "Contingency Planning Guide for Information Technology Systems," lists seven stages. It includes "Identify preventive controls" after conducting a business impact analysis (BIA) and separates plan maintenance from testing. You can download SP 800-34 from this page: *http://csrc.nist.gov/publications/PubsSPs.html*.

True or false? The BCP manager is responsible for developing and maintaining the BCP, including all elements of the plan.

Answer: *True*. The BCP manager is involved with all facets of the plan, starting with creating a project plan through coordinating BCP training, and ensuring that the BCP is regularly reviewed. However, it's also important to realize that many other personnel are responsible for their part in the plan. Typically, the BCP manager will coordinate with each department to ensure that the needs of all business entities are addressed.

EXAM TIP   BCP managers require extensive project management skills and must have the support of executive management. The creation and management of the BCP requires coordination with managers in multiple departments of the organization, and the BCP manager rarely has direct authority over any of them. An effective project manager has the skills to coordinate the efforts of the different departments, and when conflicts occur, the BCP manager has the support from executive management to resolve them. If management does not support the plan, a business case can be used to identify the benefits of the plan, the current status of recovery plans, and the potential losses without a plan.

MORE INFO   An often repeated statistic is that 93 percent of companies that lost their datacenter capabilities for 10 days or more filed for bankruptcy within one year. The validity of the statistic can't be verified, but the article "Disaster Recovery and Business Continuity Planning in Action: Japan 2011" shows how valuable effective planning can be: *http://www.microsoft.com/download/confirmation.aspx?id=26680*. Immediately after the 2011 earthquake and subsequent tsunami affecting Japan, Microsoft's network capacity was reduced to 6 percent capacity, but all services were restored to near normal capacity within three days.

True or false? The primary goal of business continuity planning is to ensure that the organization can sustain business viability after a disaster.

Answer: *True*. Business continuity planning is designed to ensure that the business can continue to function after a disaster, contributing to its long-term success. This is often referred to as ensuring the survivability of the organization.

Can you answer these questions?
You can find the answers to these questions at the end of the chapter.
1. What is defined in the initial planning phase of business continuity?
2. How does an organization identify and prioritize critical business functions and systems?
3. What are the responsibilities of the BCP manager?
4. What is the responsibility of executive management in relation to the BCP?

Objective 8.2: Conduct business impact analysis

Completing the BIA is an integral step in the creation of the BCP. It identifies the critical business functions that the plan should cover and provides a direct input into the recovery objectives.

Exam need to know...

- Identify and prioritize critical business functions
  For example: What is a critical business function? What determines the priority of a critical business function?
- Determine MTD and other criteria
  For example: What is the definition of MTD? Should an organization calculate MTDs on all business functions?
- Assess exposure to outages (e.g., local, regional, global)
  For example: What is a potential issue that should be considered for a wide area network (WAN)?
- Define recovery objectives
  For example: What is a recovery point objective (RPO)? What is a recovery time objective?

Identify and prioritize critical business functions

The first step in the BIA is to gather information about the organization, with the goal of identifying and prioritizing critical business functions. A critical business function is any function that is vital to the health of the organization. If it fails, the organization's ability to perform critical operations fails. A critical business function is sometimes referred to as a critical business process, and any function or process will typically have multiple elements.

For example, imagine a company that primarily sells products via the Internet. Critical business functions to support the company are (1) selling the products, (2) shipping the products, and (3) collecting the funds from banks processing credit card payments. If the company is unable to do any of these functions, this inability will impact the company's long-term survivability or viability.

Every minute a website is down equates to lost sales, so the company will likely consider the sales function as the top priority. Depending on the business model and the products, short delays in shipping products might be acceptable. Similarly, short delays in receiving payments from banks might be acceptable. In this scenario, the company might prioritize the three functions in this order: selling, shipping, and collecting funds.

The BIA also identifies the supporting elements of the critical functions. For example, the selling function includes multiple elements, such as Internet connectivity, a web server, probably a separate database server, and a firewall or perimeter network. If the organization wants to ensure that the website is never down, it must address each of these elements.

> **EXAM TIP** After the elements of critical business functions are identified, an organization will consider eliminating any single point of failure (SPOF) within each of these functions. This might require having alternate communication lines, failover clusters for critical servers, and one or more redundant array of independent disks (RAID) configurations for fault tolerance.

True or false? During a BIA, departments are asked to provide cost-related data to identify the impact of losing specific business-related functions.

Answer: *True.* For example, a department might be asked to identify the impact if a function is not available for 24, 48, or 72 hours. Based on the results of this data, the organization can identify and prioritize the critical business functions.

> **EXAM TIP** Two important elements that are identified for any critical business function or process are the results (or outcome) of the business function and the requirements to keep business functions operational. For example, the result of an e-commerce web server is the ability to sell products or services through the website. The requirements to keep it operational might be the web server, a separate database server, and supporting security protection.

The priority of the business functions is often based on cost-related data, but this isn't a requirement. For example, a military agency might not care about costs but instead focus on the ability to defend or attack based on its mission.

A disruption is any unplanned event that results in an outage. You can often identify critical business functions and their importance compared to other business functions by asking some key questions related to a disruption, such as the following:

- **Is this function dependent on other functions?** For example, product shipping is dependent on product sales. Without sales, there is nothing to ship, indicating that sales functions are a higher priority.
- **Are other functions dependent on this function?** For example, sales are dependent on shipping. If shipping stops completely, the company will lose the goodwill of its customers, and customers will stop buying products.
- **Will the loss of this function cause a loss of revenue? And if so, how much?** If the business function affects the company's bottom line, it is often critical. The amount of loss helps a company prioritize the function. For example, if an e-commerce website fails, a company can often identify the

loss of revenue for each hour it is down, or possibly even for each minute it is down.

- **Will the loss of this function result in noncompliance to a regulatory requirement?** Noncompliance can result in fines and lawsuits, as well as a loss of customer goodwill or a reputational loss. It often affects future sales.
- **At what time will the loss of this function affect business operations?** Choices can include 8 hours, 24 hours, 2 days, 5 days, or 10 days. The goal is to identify how critical the function is based on how long of an outage it can sustain.

The BIA is often conducted through research, a survey, and/or an interview process. For example, the BCP manager might research company documentation to identify obvious critical business functions and then send a survey to each department head with the previous questions. In some cases, the BCP manager might include some focused questions for specific departments. As the documentation is collected, the BCP manager fine-tunes the data with follow-up interviews.

Determine maximum tolerable downtime and other criteria

After the critical business functions are identified and prioritized, the organization then focuses on how long it can do without each of these functions. Some disruptions will have an immediate impact on operations. Other disruptions can be tolerated for a longer period of time.

There is a direct relationship between the MTD and the cost to recover a failed system. Functions that cannot tolerate even a moment of downtime might require expensive recovery strategies. In contrast, functions that can tolerate longer downtimes might not cost as much to protect.

True or false? The MTD identifies the maximum amount of time a critical business function can be out of service.

Answer: *True*. If the MTD is exceeded for the business function, it can cause irreparable harm to the business. This is sometimes called the maximum tolerable period of disruption (MTPOD). Business functions such as sales, billing, and customer service typically have the shortest MTD, while some human resource (HR) functions such as benefits administration will have the longest MTD.

> **EXAM TIP** It is not necessary to identify an MTD for all business functions. For example, some benefits administration services provided by the HR department might not be identified as critical business functions, and if not, it is not necessary to identify their MTD. The understanding is that as the critical business functions are restored, normal business activity can resume.

True or false? An organization will try to balance the cost of a disruption with the cost to recover from the disruption.

Answer: *True*. In general, the cost of a disruption goes up over time. The longer an outage continues, the more it impacts the business. The cost to recover a system decreases if you can tolerate the outage for a longer period of time. If you need to recover from an outage as soon as it occurs, it requires a significant investment in recovery strategies. If you can tolerate an outage of 24 hours, the recovery costs are much less.

Assess exposure to outages (e.g., local, regional, global)

The location of an organization impacts the risk related to different types of outages. For example, there is a much lower risk of a hurricane hitting San Francisco in northern California than a hurricane hitting Miami in southern Florida. Similarly, there is much higher risk of an earthquake hitting San Francisco than one hitting Miami.

True or false? An e-commerce website is susceptible to potential outages due to only local and regional issues.

Answer: *False.* An e-commerce website is publicly accessible from anyone in the world. An organization hosting an e-commerce website needs to consider risk exposure related to outages locally, regionally, and globally.

> **EXAM TIP** Local issues are primarily related to local area networks (LANs) and local connectivity. Regional issues extend to WANs and connectivity between the LANs. Alternate paths of connectivity help protect against exposure to these outages. Global issues relate to any organization selling or operating in more than one country or region.

Define recovery objectives

The recovery objectives are identified after identifying the MTDs for the critical business functions. Two important terms related to recovery objectives are RTO and RPO.

True or false? An organization has determined a critical business function has an MTD of one hour. Based on this, the organization should set an RTO of one hour or more.

Answer: *False.* The RTO should be equal to or less than the MTD.

- The RTO identifies how quickly the business function should be restored. It is always equal to or less than the MTD. For example, if the MTD for a business function is one hour, the RTO is one hour or less.
- The RPO identifies how much data loss the business can tolerate. For example, an employee database is regularly updated by HR personnel using paper records. If the database fails, losing a week of data updates might be acceptable and the updates can be re-created from the paper records. In contrast, an e-commerce website using a server database to record sales might have an RPO of one minute or less. If the database fails, the organization wants to ensure that all of the sales information up to the moment of failure can be restored.

> **EXAM TIP** The RPO is most commonly associated with databases, and it identifies how much data loss is acceptable. However, the same concepts apply to any type of backup.

Can you answer these questions?
You can find the answers to these questions at the end of this chapter.
1. When documenting a critical business function in a BIA, what are the two most important points about the business function that should be included?
2. In what stage of overall business continuity planning should the MTD of a system be identified?
3. What potential outages can impact a WAN?
4. An organization determined that data loss of up to one hour is acceptable for an online database. What have they defined?

Objective 8.3: Develop a recovery strategy

For this exam objective, the focus shifts from the big picture of business continuity to a more focused disaster recovery strategy for specific elements of critical business functions. Disaster recovery strategies are needed in the event of a disaster and focus on restoring IT systems to a functioning status.

Exam need to know...

- Implement a backup storage strategy (e.g., offsite storage, electronic vaulting, tape rotation)
  For example: What is the difference between an incremental and a differential backup? What is remote journaling?
- Recovery site strategies
  For example: What is included in a cold site? What type of site is the easiest to test?

Implement a backup storage strategy (e.g., offsite storage, electronic vaulting, tape rotation)

The primary purpose of a backup is to create copies of critical data that can be restored after an outage or disruption. There are many different types of strategies and criteria for using them.

True or false? A differential backup will back up all files that have been modified since the last full backup.

Answer: *True*. Traditional backup strategies start with a full backup and then use either incremental or differential backups. Differential backups back up all files that are new or have been modified since the last full backup, without regard for recent differential backups. An incremental backup backs up all files that are new or have been modified since the last full or the last incremental backup.

> **EXAM TIP** The MTD and RTO should be considered when deciding between using a full/incremental or full/differential backup strategy. A full/differential backup strategy can be restored more quickly because it requires only the full backup and the most recent differential backup. The drawback is that it takes longer to complete each

differential backup than the previous one. A full/incremental backup takes less time for each incremental backup but requires more time to restore because you must restore the full backup and each incremental backup created since the full backup.

True or false? Backups of sensitive data should be stored only onsite to prevent loss of confidentiality.

Answer: *False*. A core principle of backups is that a copy of a backup should be kept offsite (in a separate geographical location). If a disaster such as a fire destroys the primary location, the data can still be restored from the offsite location. The backups should have the same level of protection as the original data.

> **EXAM TIP** The best way to test the effectiveness of a backup strategy is to restore the data from a backup tape. For example, personnel can retrieve a backup from the offsite location and restore it to an offline server. This verifies that backups work and also provides training to personnel for how to perform a restore. The first time personnel perform data restoration should not be during a disaster or after a major disruption.

The primary method of storing backups has been on tapes. As disk drives have become larger and less expensive, many organizations perform backups to disk drives, but tapes are still widely used. One of the challenges with tapes is that they will wear out, so they need to be rotated out of service before they reach the end of their useful lifetime. A backup stored on a tape that cannot be read is the same as having no backup at all.

True or false? An organization using the Towers of Hanoi tape rotation scheme can keep backups for more than 30 days with only six tapes.

Answer: *True*. The Towers of Hanoi tape rotation strategy is based on the Towers of Hanoi logic puzzle. You can calculate how many days a certain number of tapes will store by using the formula 2^{n-1}, where n is the number of tapes. Six tapes is 2^{6-1}, or 2^5, which is 32 days. Another more popular tape rotation strategy that isn't as complex is the grandfather-father-son backup strategy.

True or false? If a system is using a fault-tolerant RAID such as a RAID-1 or RAID-5, backups are not needed.

Answer: *False*. Fault tolerance and backups are not the same thing. A catastrophic failure in a server can destroy all disks. Similarly, a fire can destroy all disks. If backups don't exist, the data is lost forever.

Electronic vaulting refers to sending data offsite electronically for backup purposes. Many third-party companies provide the service over the Internet, and it's also possible for an organization to configure electronic vaulting between two different locations by using a virtual private network (VPN) or other semi-private link. Electronic vaulting is typically associated with databases, but it can refer to any type of backup transmitted electronically to a remote location.

Tape vaulting is similar to electronic vaulting except that data is stored on tape at the remote facility. In contrast, electronic vaulting can have backups stored on tapes, hard drives, or other media.

MORE INFO Some companies suggest using electronic vaulting for automated backups instead of traditional backups. For example, Adaptive Data Storage (ADS) suggests using electronic vaulting to reduce costs (*http://www.adaptivedata.com/evault.htm*). Its services include encryption of the data and storage of the backups locally and remotely. Remotely stored data is further protected with offsite mirrored servers.

Remote journaling is similar to electronic vaulting except that only changes are sent to the remote facility. These changes are sent as transaction logs or journals. Like electronic vaulting, remote journaling is associated with databases.

Remote mirroring builds on remote journaling by first configuring an identical database at the remote facility. Then, all changes to the original database are sent in batches as transaction logs or journals to the remote facility and applied to the remote mirror. After the changes are applied, the remote mirror is up-to-date with the original database. Changes can be sent instantaneously as they occur or periodically, such as every 15 minutes.

EXAM TIP The differences between electronic vaulting and remote journaling are subtle and sometimes blurred. You might run across different definitions, but the following distinctions are most common. Electronic vaulting creates full backups, but after a backup is created, electronic vaulting can be configured to send only changes to the remote facility. Remote journaling never does a full backup but sends the changes only in transaction logs or journals.

MORE INFO The Business Continuity Management Institute (BCM Institute) maintains several wiki glossaries. The page at *http://www.bcmpedia.org/wiki/BCMBoK_4:_Recovery_Strategy* includes links to their definitions for several business continuity management and disaster recovery terms.

Recovery site strategies

A recovery site is an alternate location that an organization can use in the event of a disaster. For example, if a fire, earthquake, hurricane, or flood occurs, the organization can move critical business functions to the alternate location. These alternate sites are often referred to as continuity of operations plan (COOP) sites.

True or false? Hot sites provide an alternate location that can be brought into service in a very short period of time.

Answer: *True*. Hot sites are alternate locations that can be brought into service in the shortest amount of time compared with other recovery sites.

Common recovery sites are cold sites, hot sites, and warm sites.

- **Cold site** This is a building with electricity, running water, and communication links. It typically doesn't include any hardware or data, but in some cases, it has boxed up hardware that must be configured when the cold site is implemented. Cold sites are the least expensive but the most difficult to test, and they take the longest to become operational.

- **Hot site** This location has all the hardware in place and is fully configured. It includes up-to-date data that is often kept up to date with real-time synchronization. It's common for a hot site to be manned with personnel performing other business functions, and additional personnel come to the site when it is implemented. Hot sites can become operational within hours. They are the most expensive and take the least amount of time to become operational.
- **Warm site** This is a compromise between a hot site and a cold site. Warm sites will typically have the hardware configured but not powered on. The hardware might have some data, but it will need to be updated.

EXAM TIP The actual time required to switch over to an alternate site isn't easily defined. In general, a hot site has a very short switch-over time and can be measured in hours, such as one to four hours. A cold site has a long switch-over time and can be measured in days, such as two to three days. A warm site has a medium switch-over time and is somewhere in the middle.

The following two types of alternate sites are referenced less often, but you should have a basic understanding of them:

- **Mobile site** This is a self-contained, transportable container that includes the equipment necessary to provide specific services. For example, a container typically hauled by tractor-trailers can be outfitted with telecommunication and IT equipment and transported to any location.
- **Mirrored site** A mirrored site is a step above a hot site. It is a fully redundant site that is identical to the primary site. Backup strategies such as remote mirroring are used to ensure that the mirrored site always has the same data as the primary site.

A less-used option is a reciprocal agreement where one company agrees to let another company use its facilities during a disaster. While it's cheaper, it isn't as reliable. The alternate company might have limited resources during the disaster. Additionally, changes at the alternate location might make it impossible to use their facilities unless both organizations regularly exchange change management information.

If an organization is using an alternate location, it should have clear steps for how to implement a disaster recovery plan and move critical business functions to this location. These steps should include the priority of the systems and ensure that systems are restored in the correct order based on inter-dependencies between the systems.

EXAM TIP No matter what type of site is used, it's imperative that the site has the ability to access the data. For example, the primary location might regularly back up data to tape and store copies of the tape in an offsite location. If the tape backup system at the primary location is updated, it's possible for the business data to be inaccessible at an alternate location unless it has a compatible tape system to read the new tapes.

MORE INFO Many organizations are using virtual and cloud-based solutions rather than maintaining a physical site. This can often lower costs in addition to reducing downtime of critical business functions. VMware published an article titled "Improving Business Continuity with VMware Virtualization" that identifies many common challenges with business continuity and some of the inherent benefits. You can access it here: *http://www.vmware.com/files/pdf/VMware-IMPROVING-BUS-CONT-SB-EN.pdf*.

When an organization moves to an alternate location, it should establish a clear line of authority at this location. It's very likely that all personnel will not move to the alternate location, so the plan will identify who is in charge. Additionally, the plan will include an order of succession. For example, if a specific department head is expected to run the alternate location but for some reason is not there, the plan should identify the next person who should assume authority. The order of succession is commonly listed by job title rather than by name.

Can you answer these questions?

You can find the answers to these questions at the end of the chapter.

1. An organization uses a full/differential backup. It does a full backup on Sunday and differential backups on every other day of the week. If a system fails on Wednesday, how many backup tapes are required to restore the data?
2. An organization wants to maintain a copy of all changes to an online database at a remote location. The changes should be sent to the remote location as they occur. What should the company use?
3. What type of alternate site location is the most difficult to test?
4. What type of alternate site location receives changes from the primary site as they are occurring?

Objective 8.4: Understand disaster recovery process

For this exam objective, you need to have an understanding of the overall disaster recovery process. That is, when a disaster is declared by someone on the disaster recovery team, the DRP provides the steps to restore critical business functions.

Exam need to know...

- Response
  For example: What is a reliable method of notifying personnel?
- Personnel
  For example: Is it possible to have more than one technical recovery team? When a disaster occurs, what should be valued the most?

- Communications
  For example: Who should be notified when a disaster occurs? What can be used to prevent someone from falsely declaring a disaster?
- Assessment
  For example: What is the responsibility of the damage assessment team?
- Restoration
  For example: What will be operational at the end of the restoration phase? What is the difference between restoration and reconstitution?
- Provide training
  For example: Do personnel receive the same type of DRP training?

Response

When a disaster strikes, the disaster recovery team responds and begins to implement the DRP. Some of the steps in this process include the following:

- Notify all personnel on the DRP teams.
- Notify customers and business partners of the outage and the expected time when the system will be recovered.
- Restore critical servers. Activate alternate locations if necessary.
- Retrieve backups from either onsite or offsite locations and begin restoring data.
- Test systems for critical business functionality and to ensure that systems are secure.
- Restore network functionality.

A key part of a disaster recovery response is the notification of personnel. Depending on the type of disaster, personnel can be notified immediately after the disaster or when the disaster is imminent. For example, personnel are notified immediately after a fire or earthquake. If a hurricane is approaching, personnel are notified ahead of time, such as 24, 48, or 72 hours prior to when the hurricane is expected.

Personnel that should be notified include executive management and all personnel involved in disaster recovery actions. A DRP will identify personnel's responsibilities when a disaster occurs, and each person will respond based on their assigned responsibilities.

True or false? A reliable method of notification is the use of a call tree.

Answer: *True*. A call tree lists all personnel that should be notified. Phone calls provide confirmation that everyone that should be notified has been notified.

> **EXAM TIP** Notification procedures should be well documented in the DRP, including both primary and alternate contact methods. Notification methods include email, phones with call trees, personal contact, and handheld radios.

Personnel

Personnel involved in a disaster recovery plan are typically divided into teams. Some of the common teams include the following:

- **Emergency management team** This includes the business continuity plan coordinator and senior management. This team can include executive personnel, but if not, members of this team should have direct access to executive management to communicate the status of the incident.
- **Damage assessment team** Members of this team assess the damage after a disruption. Their goal is to quickly identify the extent of the damage and the impact it will have on the organization's mission.
- **Technical recovery teams** Members of these teams are responsible for restoring systems after a disruption, and they are sometimes referred to as salvage teams. Depending on the functions that need to be restored, an organization might have just one or two technical recovery teams or it might have multiple teams.

EXAM TIP An organization should always value people more than things. With this in mind, personnel safety should always be a primary concern. For example, if a fire occurs, immediate evacuation of personnel should be the first priority. When considering electronic door locks, personnel should be able to open them to escape a fire even if power is not available.

Communications

Before, during, and after a disaster, personnel need to have a clear understanding of what communication methods to use. This includes primary and alternate methods of communication. It's also important that personnel have a clear understanding of who to communicate with, including employees, customers, and the media. Typically, only designated personnel are authorized to communicate with the media, and all personnel should be aware of this to prevent problems resulting from miscommunication.

During major disasters, it is common to set up a "war room" for centralized communications. For example, an organization can designate a conference room where personnel have periodic meetings to discuss progress and also where everyone can report major milestones. Another example might be a technical recovery team reporting when a major system is operational.

True or false? A DRP should identify who can declare a disaster and include authentication codes used by these individuals.

Answer: *False*. A DRP should identify who is authorized to declare a disaster, and authentication procedures should be in place to prevent unauthorized individuals from declaring a disaster. However, if authentication codes are published in the DRP, they are widely available.

True or false? It is not recommended to let customers know via a website that services are limited because an organization is implementing a DRP.

Answer: *False.* Communicating to customers about a disaster and that the organization is taking preplanned steps to respond is recommended. It presents a favorable view of the organization and communicates to customers that the organization is prepared. When a company just stops providing services without explaining why, it looks unprepared.

Members of disaster recovery teams will need to communicate with each other throughout the disaster. Many times one team will be recovering a system that is necessary for another team. For example, one team might be restoring Internet capabilities while another team is restoring a website. Both teams must succeed before the website is operational again.

Assessment

After an incident occurs, members of the damage assessment team investigate to determine the extent of the damage. In some cases, the result of this assessment can result in the organization declaring a disaster. In other cases, the assessment occurs after the disaster has been declared.

For example, after an earthquake hits, the team assesses the damage. Based on the results, the organization can decide to declare the disaster and move operations to an alternate location or continue to operate where they are.

In contrast, an organization can decide to declare a disaster when a damaging hurricane is within 24 hours of striking. Before the hurricane hits, the organization moves business functions to an alternate location. After the hurricane hits, the damage assessment team determines the extent of the damage and the organization can identify when it can begin normalizing business functions.

True or false? A damage assessment team can declare a disaster and authorize movement of critical business functions to an alternate location.

Answer: *False.* A damage assessment team identifies the extent of the damage, and they are responsible for reporting the damage. However, they are rarely authorized to declare the disaster. Instead, the DRP identifies specifically who is authorized to declare the disaster or what events cause the disaster to be declared.

> **EXAM TIP** Assessments should be completed as quickly as possible but not at the expense of the safety of personnel on the assessment team. Assessments identify current damage, potential for further damage, and the impact on critical business functions.

Restoration

Restoration refers to taking the necessary steps to ensure that critical operations are returned to service. It could include implementing an alternate site or taking the necessary steps to repair damage and restore a system at the primary location.

True or false? Restoration refers only to restoring servers to operation.

Answer: *False*. Restoration refers to restoring any critical business function or system. It includes any critical desktop and portable systems, servers, mainframe systems, LANs, WANs, websites, and alternate sites. It does not include noncritical operations.

A key element of successful restoration is a clear understanding by the recovery team of what to recover and how to do so. The team should be very familiar with the steps to recover a system and have alternate methods of retrieving these steps. For example, if the recovery steps are stored on a server that is no longer available, the team should either have access to the steps from another location or be familiar enough with the steps that they can perform them without having the files.

Teams should have a clear idea of the priority of system recovery whenever they are responsible for recovering multiple systems. This is often driven by the information gathered from the BIA but might also be driven based on the requirements for a system. For example, a web server might need a database server to be operational before it will operate.

> **EXAM TIP** At the completion of the recovery phase for any single system, the system will be operational and performing its dedicated functions. This might require the recovery team to implement contingency plans so that the system might not be operating at the original location or with all of its redundant capabilities, but it will be performing critical business functions.

> **MORE INFO** NIST SP 800-34 stresses the importance of ensuring that recovery procedures are straightforward and written in a step-by-step style. It includes examples for simple recovery process checklists and contingency plans for critical systems.

The reconstitution (or reconstruction) phase occurs after the recovery phase and will end all recovery activities. At this point, normal operations are transferred back to the original location if an alternate location was activated. All primary systems are fully restored, and any contingency measures that were implemented to restore critical business functions are no longer needed.

When transferring functions from an alternate location, the least critical functions should be transferred first. There are sure to be some minor problems that occur during this process, and if the least critical functions are transferred first, they are the ones that are affected rather than the most critical functions. By the time the critical functions are ready to be transferred, all the minor problems should be identified and resolved.

> **EXAM TIP** After an incident, the organization should complete a "lessons learned" review and evaluate the usefulness of the plan. It's likely that some things worked well, whereas others can be improved. This review allows the organization to implement changes to better meet the needs of the organization if the BCP needs to be activated again.

Provide training

From a big picture perspective, training helps employees within an organization understand that security is everyone's responsibility. DRP and BCP training provides details to all personnel on their responsibilities within the plan and how they can fulfill their responsibilities.

True or false? One of the primary goals of DRP/BCP training is to ensure that employees understand their responsibilities.

Answer: *True*. Employees need to understand their responsibilities and their role in the DRP/BCP.

Training is tailored to individuals and groups as necessary. For example, technicians and administrators need technical training to perform specific tasks in support of the DRP and BCP. Management needs different training, such as understanding what constitutes a disaster and when disaster recovery plans should be implemented.

Can you answer these questions?

You can find the answers to these questions at the end of the chapter.

1. Who should be notified when a DRP is activated?
2. What are some common teams identified in a DRP?
3. When planning for communication during a disaster, what should be a primary consideration?
4. When members of the assessment team are performing assessments, how much time do they spend restoring systems?
5. How do members of the disaster recovery teams know which systems to restore?
6. What type of training is provided to personnel on disaster recovery teams?

Objective 8.5: Exercise, assess, and maintain the plan (e.g., version control, distribution)

For this exam objective, it's important to realize that the BCP is rarely created perfectly in the first draft. Instead, exercises and reviews are completed to ensure that the BCP meets the intended goals. There are a variety of different exercises that can test the plan on different levels, and any of these can result in a modification to the original BCP. Version control procedures ensure that all personnel know what the current plan includes and what changes have been made to the plan.

Exam need to know...

- Exercises
  For example: What is a tabletop exercise? Do exercises simulate actual disasters?

- Maintain the plan
  For example: How often should the BCP be updated? How are changes to the BCP documented?

Exercises

Exercises are used to test different elements of the plan. Multiple types of exercises are used, and initial exercises have the least impact on operations.

Typically, all departments will review the BCP individually prior to any formal testing and ensure that it meets their needs. After their changes are implemented, formal exercises are started to fine-tune the plan. Some common types of exercises include the following:

- **Tabletop exercises** Participants meet and talk through the process of a scenario identified in the BCP in a tabletop exercise (or structured walk-through exercise). It's commonly done with participants sitting around a conference table or in a training room. For example, the BCP coordinator might state a hurricane is 72, 48, 24, and then 12 hours away. At each phase, participants state their response. A tabletop exercise can often identify logical flaws in the plan.
- **Functional exercises** These exercises test individual elements of a plan. For example, it can test the ability to retrieve offsite backups and restore them to a critical server in a cold site. As another example, the organization can test the ability to rebuild a critical server just as it would need to do if the critical server failed.
- **Parallel tests** These tests are used for organizations that have alternate facilities. Specific systems or functions are activated at the alternate location while keeping the primary location running without interruption. Systems at the alternate location are evaluated to determine whether they are functioning at the same level of service as the primary location.
- **Full-scale exercises** These provide a more realistic test of the plan and simulates an actual disruption of critical business functions. They are the most expensive to perform and should be done only after tabletop and functional exercises have been completed.

True or false? A functional exercise performs the actual steps of each specific element of a BCP.

Answer: *True*. A functional exercise tests the individual elements of DRPs and the BCP. It can be as simple as exercising a recall roster or as complex as testing an alternate recovery site.

True or false? A disaster recovery test must simulate actual disaster conditions.

Answer: *False*. Tabletop exercises, walkthroughs, and functional exercises do not simulate the actual disaster conditions. Participants are often asked to consider the factors they expect to experience during a disaster, but these factors don't need to be simulated. Any testing that simulates actual disaster conditions can affect normal operations and should be minimized. The disaster recovery team identifies how real the tests should be.

True or false? The BCP and DRPs need to be tested only if an organization does not have a hot site.

Answer: *False*. All plans need to be regularly tested. Having a hot site doesn't negate the need for testing the BCP or DRPs.

> **EXAM TIP** The primary purpose of exercises is to verify that the plans can meet the goal of keeping critical business functions operational after or during a disruption.

True or false? One of the most important outcomes of an exercise is the documentation.

Answer: *True*. The documentation identifies whether the plan is effective or has issues. Between testing of any of the plans, steps are taken to correct any of the issues or flaws identified in previous tests.

> **EXAM TIP** Personnel involved in exercises and testing are responsible for correcting any discovered flaws. Often this requires rewriting procedures and assessing the impact of new controls.

Maintain the plan

A BCP is a living document and should be regularly updated to meet the changes in the environment and the organization. There are several reasons why a BCP can become out of date, including growth within the organization, new acquisitions, changes in system configurations, changes in the building or environment, and personnel changes. If these changes affect the usefulness of the BCP, it needs to be updated to reflect the changes.

True or false? Personnel responsible for the BCP should be a part of the change management process.

Answer: *True*. Changes in the organization can affect the BCP. If the BCP team is part of the change management process, they will be able to evaluate the changes to determine whether the BCP needs to be modified as a result of a change.

A BCP should include version control data so that all personnel can verify that they have the current version and easily identify the changes made to the BCP. This can be a simple page at the beginning of the document listing each change, the page number, the date it was made, a brief description of the change, and a signature by the person making or authorizing the change.

> **EXAM TIP** A BCP should be reviewed regularly to ensure that it still meets the needs of the organization. A large organization that regularly acquires new companies might review it once a quarter, while a smaller organization might review it once a year. The BCP might be updated to implement lessons learned after an incident, after a significant change in the organization, or after a review. When the BCP is updated, a copy of the updated BCP should be distributed to all personnel responsible for BCP planning and implementation.

Can you answer these questions?

You can find the answers to these questions at the end of the chapter.

1. What is the impact on operations during a functional test?
2. What should be accomplished prior to performing a full-scale exercise?
3. How should changes be documented in a BCP?

Answers

This section contains the answers to the "Can You Answer These Questions?" sections in this chapter.

Objective 8.1: Understand business continuity requirements

1. The scope of the project and the initial project plan are defined in the initial planning phase of business continuity.
2. An organization completes a BIA to identify and prioritize critical business functions.
3. The BCP manager is responsible for creation, development, and maintenance of the BCP plan.
4. Executive management must understand the plan, support it, and support the BCP manager.

Objective 8.2: Conduct business impact analysis

1. The BIA should include the result of the critical business function or the outcome it provides to the organization. Second, it should include the requirements to keep it operational.
2. The MTD is identified during the business impact analysis.
3. Connectivity outages between the networks in the WAN are a primary concern. Redundant connections using alternate paths provide protection against these outages.
4. They have defined the RPO. The RPO defines how much data loss is acceptable.

Objective 8.3: Develop a recovery strategy

1. Two backup tapes are needed: the full and the most recent differential backup. The most recent differential backup holds all the changes since the last full backup.
2. Remote journaling would meet this need. Changes are sent to the remote location in a journal or transaction log.

3. Cold sites are the most difficult to test because none of the equipment is configured and none of the data is up to date. The benefit is that a cold site is the least expensive.
4. Both a hot site and a mirrored site receive changes as they are occurring.

Objective 8.4: Understand disaster recovery process

1. All personnel identified in the DRP should be notified when a DRP is activated. This includes employees, members of disaster recovery teams, customers, and business partners.
2. Common teams identified in a DRP include the emergency management team, the damage assessment team, and technical recovery teams.
3. A primary consideration would be secondary communication methods such as handheld radios. Primary communication methods might not be available during a disaster.
4. The assessment team should not spend any time restoring systems. Their primary responsibility is to assess the current damage and the potential for future damage. They need to report this information back to the executive team as soon as possible so that the executive team can make informed decisions.
5. The DRP identifies the priorities of systems and which systems should be restored first. Team members should be very familiar with the plan's priorities and how to implement them.
6. The type of training provided to DRP teams varies. Some training is general and applies to all personnel, but more often, members of the teams have targeted training based on their specific responsibilities.

Objective 8.5: Exercise, assess and maintain the plan (e.g., version control, distribution)

1. A functional exercise will not impact operations. It tests a specific element of a plan but is isolated from production systems.
2. The plan should be tested using less intrusive methods, such as a tabletop exercise prior to performing a full-scale exercise. A full-scale exercise can impact operations, and this impact should be minimized.
3. Some type of version control should be implemented so that the changes are authorized, easily recognizable, and understood.

CHAPTER 9

Legal, regulations, investigations, and compliance

The Legal, Regulations, Investigations, and Compliance domain is focused on legal and regulatory requirements and ethical behavior. It includes forensic procedures and other topics used to support security incident investigations and other types of computer crime. This domain also addresses some of the compliance requirements that should be considered within information technology (IT) security.

This chapter covers the following objectives:

- Objective 9.1: Understand legal issues that pertain to information security internationally
- Objective 9.2: Understand professional ethics
- Objective 9.3: Understand and support investigations
- Objective 9.4: Understand forensic procedures
- Objective 9.5: Understand compliance requirements and procedures
- Objective 9.6: Ensure security in contractual agreements and procurement processes (e.g., cloud computing, outsourcing, vendor governance)

Objective 9.1: Understand legal issues that pertain to information security internationally

This domain starts with some generic information about computer crime and intellectual property and then moves into detailed issues related to international IT security. It is important to recognize that there are different types of computer crime, along with different types of intellectual property, even if an organization operates in a single country/region. However, when organizations start working internationally, there are more concerns. Organizations exporting goods and services outside the United States must follow different regulations, depending on the receiving country/region. Regulations can become extremely complex when data transactions cross a border, but there are some guidelines to simplify things. Last, privacy issues are important to consider in almost every region around the world.

217

Exam need to know...

- Computer crime
  For example: What are the three categories of computer crime? What category of crime is a salami attack?
- Licensing and intellectual property (e.g., copyright, trademark)
  For example: What are the different categories of intellectual property? What is the difference between a copyright and a trademark?
- Import/Export
  For example: Is it legal to export high-performance computers to any country/region from the United States?
- Trans-border data flow
  For example: What is trans-border data flow? What is the safe harbor framework?
- Privacy
  For example: What is the relationship between privacy and personally identifiable information (PII)? What is identity theft?

Computer crime

When people think of computer crime, they often think only of attacks on computers. However, from a legal perspective, computer crime is often divided into separate categories. This includes crimes that involve computers or networks where the computer is used to assist in the crime, crimes where the computer is the target of the crime, and crimes where the computer is incidental to the crime.

True or false? The storage of child pornography on a computer is an example of a computer-assisted crime.

Answer: *False.* Storing child pornography on a computer is recognized as a computer crime where the computer is incidental to the crime.

The following list provides detailed information about the computer crime categories:

- **Computer as the instrument (sometimes called computer-assisted or computer as a weapon)** This category includes computer crimes where criminals use computers as a tool or weapon to break the law. Typically, criminals are using computers due to the speed and accurate repetitive processes of the computers, such as in salami attacks, click fraud, or other fraudulent attacks attempting to steal funds. It's common for criminals to use this type of computer crime for monetary gain.

    NOTE In a salami attack, the attacks take a very small amount of money from a large number of transactions. For example, they might take one tenth of a penny from ten million transactions. The percentage of the penny might not be missed, but when multiplied by 10 million transactions, it results in a theft of US$10,000. The name salami attack is a reference to thin slices of salami used in sandwiches.

In click fraud, criminals use software to simulate clicks on advertisements. Criminals receive funds for each simulated click even though a customer isn't actually clicking on an advertisement.

- **Computer is the target (computer-targeted)** This category includes any crimes where the criminal identifies a specific computer to attack. It includes Denial of service (DoS) and distributed DoS (DDoS) attacks and also any unauthorized attempt to access, modify, or delete data. There are many reasons why criminals might attack, but the key here is that when a computer is attacked, it is a computer-targeted crime.

- **Computer is incidental (sometimes called computer as an accessory)** This category includes any crimes where the computer is not essential for a crime but is related to the criminal act. Storing or distributing child pornography with a computer is an example. Crimes in this category could be committed without the computer, but because the computer is available, it's used. For example, pornographers could store and distribute pictures with or without a computer, but they use computers because they are available.

EXAM TIP Know the three primary categories of computer crime. Much of IT security is focused on computer-targeted crime, but for this domain, it's important to be aware of the other types of crime and to be able to identify the category of a specific crime.

True or false? Attacks by an Advanced Persistent Threat (APT) are an example of a computer-targeted crime.

Answer: *True*. APTs target a specific organization and persistently attack until they reach their stated goal.

MORE INFO Chapter 1, "Access control," mentions APTs in the context of access aggregation in Objective 1.2. An APT is typically a team of attackers sponsored by a government, and they have the resources to patiently and persistently attack a specific target until they achieve their objectives.

Licensing and intellectual property (e.g., copyright, trademark)

Organizations often expend a significant amount of time and money to create products that they sell. Many of their products are identified as intellectual property and protected under laws throughout the world. The primary legal issue with intellectual property is that criminals steal the products without paying for them, resulting in losses to the organization that created them. Intellectual property includes trademarks, copyrights, patents, and trade secrets.

True or false? A company can trademark a logo to protect it as intellectual property.

Answer: *True*. A trademark provides legal protection for a logo and gives a company legal recourse to prevent another entity from using the logo after it has been trademarked.

Trademarks provide protection for a word, name, shape, or any combination of these when used as a unique identifier for product, service, or business. Unregistered trademarks are identified with the letters TM, and registered trademarks are identified with a circle around the letter R (®). In most cases, a trademark needs to be registered with a government entity to be fully protected. However, an organization still has legal protection against the unauthorized use of an unregistered trademark. There isn't a time limit on trademarks as long they remain in use. However, if an organization stops using a trademark for a specific time period (such as five years in some countries/regions), it loses the rights to the trademark.

True or false? Intellectual property such as books and music must be patented to be protected against piracy.

Answer: *False*. A copyright (not a patent) protects creative works such as books (or any original writing), music compositions and recordings, software applications, artwork, and more. Patents are used for inventions and grant the owner of the patent exclusive rights to the invention.

Copyrights provide protection for the creator of an original work and prevent others from legally using the work without permission. A copyright is identified with a circle around the letter C (©), the word Copyright, or the abbreviation Copr. Copyrights have definitive time frames, although they vary in different countries/regions.

True or false? The specific method an organization uses to streamline a business process to provide superior service to customers can be copyrighted.

Answer: *False*. Processes and methods cannot be copyrighted.

An organization would typically protect methods or processes as a trade secret to keep them from their competitors. There is no time limit on a trade secret.

The laws governing intellectual property vary from region to region, but most countries/regions respect the value of intellectual property. Still, the amount of losses from intellectual property is staggering. The United States Federal Bureau of Investigation reports that U.S. businesses lose billions of dollars a year from intellectual property theft. This translates to lost jobs and lost tax revenues

> **EXAM TIP** Organizations need to take precautions to protect their intellectual property. Failure to do so can result in one organization spending all its money on research and development and a criminal stealing the information and benefiting without doing any research or development. Copyrights protect creative works such as software, music, and books. Trademarks protect logos or words used as a brand. Patents protect inventions.

MORE INFO The U.S. FBI maintains an Intellectual Property Theft website here: *http://www.fbi.gov/about-us/investigate/white_collar/ipr*. Microsoft maintains a group of webpages discussing intellectual property, copyrights, trademarks, and piracy. These pages show how one company informs readers about their rights and also provide many examples of the different types of intellectual property. You can view them here: *http://www.microsoft.com/about/legal/en/us/IntellectualProperty/*.

True or false? Shareware is free to use without an obligation to purchase a license to use it.

Answer: *False*. Shareware is free to try, but the user has an obligation to pay for the software if they choose to use it beyond a specific trial period.

Software licensing includes the following categories:

- Commercial software comprises full versions of software sold to users. Software organizations generate revenue primarily through the sales of commercial software.
- Academic software is reduced-price software sold to students and faculty. It will often have the same capabilities as the full version but is provided at a reduced cost.
- Trial software (sometimes called shareware) is software that is free to use for a trial period. After the trial period is over, the user is legally obligated to stop using the software or pay to continue using it. Some trial software has limited features or will stop working when the time period ends.
- Freeware is free to use without any obligation to pay for it. Authors of freeware sometimes ask for donations, but donating isn't a requirement.
- Open source licenses specify that the source code will be freely available for others to use.

EXAM TIP Almost all software includes a license to use it that specifies how the software can be used and distributed. Different categories have different pricing.

MORE INFO Microsoft often releases trial versions of new software applications and operating systems that you can download and install for free. Go to their download site (*http://www.microsoft.com/en-us/download/default.aspx*), and search "trial downloads". Microsoft also has several academic software programs in place to support educators and students with special discounts. The following page shows some of the programs: *http://www.microsoft.com/education/en-us/buy/Pages/academicsavings.aspx*.

The key points related to software and intellectual property can be summarized as follows:

- **Copyright** Most software comes with a copyright, which provides protection for the developer or organization that produced it. Open source software providers use copyright laws to enforce the license terms but not to protect the code, and they refer to it as *copyleft*.

- **Patent** Software patents have been granted in some countries/regions, such as in the United States, but are prohibited in other areas, such as in the European Union (EU).
- **Trade secret** A trade secret isn't registered like a copyright or patent. Instead, the knowledge is kept secret within the organization.

Software often includes an end user license agreement (EULA). The EULA states how the software can be used, including any restrictions on the copying or distribution of the software. When the user agrees to the EULA, the user is bound by terms of the agreement as a contract. Some of the common contract terms include the following:

- **Shrinkwrap contract** This refers to a software license agreement included within a software package. The customer is bound to the agreement after removing the shrinkwrap covering and opening the software box. Often, the contract will be printed on paper and available inside the box, and the software media will be sealed, with an indication that breaking the seal binds the user to the agreement.
- **Clickwrap contract** These are used for online transactions after users download and install software. Users are forced to click a button indicating that they have read and agree to the contract. If they choose not to agree to it, they simply don't install the software.
- **Browsewrap contract** Many websites have specific terms of use that apply when users visit and use the website. These have sometimes been determined to be unenforceable because they weren't easily accessible.

EXAM TIP Organizations that attempt to hide or not fully show contracts often find that the contract is not enforceable in a court of law. Contracts should be readily apparent to any user so that they are aware that they are entering into a contract. This is commonly done with most shrinkwrap and clickwrap contracts. However, many browsewrap contracts aren't as clear and might be challenged in courtrooms.

Import/Export

With the widespread use of the Internet around the world, it's become relatively easy for an organization to import and export goods and services. Some products aren't available locally, so they must be imported. Other products seem classier when they're imported, and some are simply cheaper when purchased from another country/region. All of these factors contribute to more imports and exports.

True or false? When exporting goods and services, an organization needs to be concerned only with the country/region of origin.

Answer: *False*. An organization that exports goods or services must abide by the laws of the country/region of origin and the country/region to which the goods are being exported. In some cases, the organization also has to concern itself with any countries/regions through which the goods are transferred.

The exporting of high performance computers (HPCs), software, and encryption products is regulated in the United States. Specifically, it is illegal to export these commodities from the United States to certain countries.

It is legal to export any commodities to countries listed as a Tier 1 country by the Bureau of Industry and Security within the U.S. Department of Commerce, but other countries have restrictions. Organizations exporting IT goods and services from the United States to non-Tier 1 countries need to take extra steps to ensure that they are in compliance with existing laws.

> **MORE INFO** You can view a list of countries in Tiers 1 and 3 on the following page maintained by the Bureau of Industry and Security of the U.S. Department of Commerce: *http://www.bis.doc.gov/hpcs/countrytier.htm.*

Trans-border data flow

Trans-border data flow refers to the movement of electronic data across government boundaries, such as across different states, countries, or regions. During these transactions, the transmission of the data is governed by different entities, which can cause legal conflicts.

True or false? Organizations that transfer data across international boundaries are required to comply only with their internal laws and regulations.

Answer: *False*. Organizations that transfer data across international boundaries must comply with regulations for each country/region.

Some international organizations have been created to help ensure that data is adequately protected while also ensuring that each country/region follows the relevant rules and regulations. The Organisation for Economic Co-operation and Development (OECD) provides guidelines for ensuring that data is protected when it is shared between countries/regions.

> **EXAM TIP** Any transaction that results in data crossing a regional boundary is considered a trans-border data flow transaction. The transaction must comply with regulations of each region. This can be quite complex and is simplified by using guidelines from the OECD.

> **MORE INFO** You can view the guidelines published by OECD on the protection of privacy and trans-border data flow here:
> *http://www.oecd.org/internet/interneteconomy
> /oecdguidelinesontheprotectionofprivacyandtransborderflowsofpersonaldata.htm.*

True or false? The Safe Harbor framework provides a method that allows countries/regions to share data while also complying with all applicable laws during trans-border transactions.

Answer: *True*. The Safe Harbor framework provides data on how U.S. organizations can comply with EU and Switzerland regulations.

The Safe Harbor framework helps ensure that all data is protected in accordance with applicable laws. This includes any personal privacy data and personal health information. One Safe Harbor framework is focused on compliance between the U.S. and the EU, and another Safe Harbor framework is focused on compliance between the U.S. and Switzerland.

> **MORE INFO** Information about the Safe Harbor frameworks is available on the U.S. Export.gov website: *http://export.gov/safeharbor/*.

Privacy

The protection of privacy data is governed by many different laws throughout the world. PII is information that can uniquely identify an individual or can be combined with other data to uniquely identify an individual. It includes elements such as the individual's name combined with the following information:

- Identification number (such as a Social Security number used in the United States or the National Insurance number used in the United Kingdom)
- A driver's license number
- Date and place of birth
- Biometric data such as from a fingerprint or DNA
- Mother's maiden name

True or false? The availability of personal information increases the incidence of identity theft.

Answer: *True*. When criminals have access to personal information, they can steal identities and use them for illegal purposes.

One of the biggest risks to individuals when their privacy data isn't protected is identity theft. Criminals use their information to create shadow identities, open bank accounts, obtain credit, and more. The criminal obtains as much money or merchandise as possible from the stolen identity through unauthorized charges. Victims often don't know about the crime until they apply for credit and realize that their credit rating has been destroyed. In other cases, they begin receiving bills for services or merchandize they never purchased.

> **EXAM TIP** Organizations often collect PII about their employees and customers, and they have a responsibility to protect all PII. Many laws require organizations to report the compromise of any PII to law enforcement organizations and also to all individuals that might have had their personal data stolen. In some cases, the breached organization offers to pay for credit monitoring services. These services help individuals determine when their credit report is having activity that they haven't generated.

> **MORE INFO** Chapter 3, "Information security governance & risk management," presents information about PII in the context of privacy requirements compliance in Objective 3.2. It also refers to various laws such as the United States Privacy Act of 2005 and the European Union Directive 2002/58/EC (the E-Privacy Directive).

True or false? The Children's Online Privacy Protection Act (COPPA) restricts collection of privacy information for children under the age of 13.

Answer: *True.* COPPA includes specific restrictions on the collection of any information for children under 13.

If a website plans on collecting information from children under the age of 13, the website owner must gain consent from a parent or guardian. Additionally, the website owner needs to add information to their privacy policy detailing what type of information they collect and how it is used.

True or false? Privacy policies used within an organization should address all governing regulations that apply to the organization.

Answer: *True.* Privacy policies should comply with all laws and regulations that apply to the organization.

Privacy policies should also take the organization's business objectives into consideration. As with any policy, the goal is to provide direction while also ensuring that the business can successfully pursue its goals and mission. An effective privacy policy can protect PII and comply with the laws without stopping the mission.

Can you answer these questions?

You can find the answers to these questions at the end of this chapter.

1. What category of computer crime is click fraud?
2. A user can download a free copy of an application but can use it for only 30 days. What type of software licensing category is being used?
3. What type of contract applies when a user visits a website that has a terms of use contract?
4. What is the primary goal of the Organisation for Economic Co-operation and Development (OECD)?
5. What is identity theft?

Objective 9.2: Understand professional ethics

Professional ethics identify the minimum rules of conduct or behavior for a professional. When written out, they provide individuals with a moral compass that they can use when resolving ethical dilemmas. For this objective, you need to have specific knowledge about the Code of Ethics published by (ISC)2, including the preamble and the canons. You also need to have a basic understanding of employees' responsibility to support the organization that employs them and their responsibility to support the organizational code of ethics.

Exam need to know...

- (ISC)² Code of Professional Ethics
  For example: What are the CISSP requirements related to the Code of Ethics? What are the four canons of the Code of Ethics?

- Support organization's code of ethics
  For example: When should an employee report ethical violations outside the organization?

(ISC)² Code of Professional Ethics

(ISC)2 has created and published its Code of Ethics, which applies to all individuals seeking an (ISC)2 certification and (ISC)2 members. The Code of Ethics includes a preamble and four canons. The Preamble is as follows:

- Safety of the commonwealth, duty to our principals, and to each other requires that we adhere, and be seen to adhere, to the highest ethical standards of behavior.
- Therefore, strict adherence to this Code is a condition of certification.

True or false? Individuals can have their CISSP certification revoked if they violate any one of the four canons.

Answer: *True*. Individuals who violate the (ISC)2 Code of Ethics, including the preamble and the four mandatory canons, can have their certification revoked.

> **EXAM TIP** All CISSP candidates must agree to the Code of Ethics before they will be awarded the certification. Additionally, any individual who has earned the CISSP certification and who intentionally or knowingly violates any provision of the Code of Ethics can have their CISSP certification revoked.

True or false? An individual with the CISSP certification is responsible for maintaining the safety of society, the commonwealth, and the infrastructure.

Answer: *True*. This is the first canon in the Code of Ethics.

The Code of Ethics canons are as follows:

- Protect society, the commonwealth, and the infrastructure.
- Act honorably, honestly, justly, responsibly, and legally.
- Provide diligent and competent service to principals.
- Advance and protect the profession.

True or false? When faced with a conflict between different canons in the in the (ISC)2 Code of Ethics, an individual should give preference to the canon appearing earlier than the conflicting canon.

Answer: *True*. According to (ISC)2, conflicts should be resolved in the order of the canons.

> **EXAM TIP** When preparing for the CISSP exam, ensure that you know the preamble, the four canons, and the order of the four canons. Adherence to the preamble and canons is mandatory, and the order of the canons should be considered if a situation presents a conflict between any of the canons.

> **MORE INFO** You can view the overview of the (ISC)² Code of Ethics here: https://www.isc2.org/ethics/default.aspx. Also, you can download a full file that provides details of the canons here: https://www.isc2.org/uploadedFiles/(ISC)2 _Public_Content/Code_of_ethics/ISC2-Code-of-Ethics.pdf.

Support organization's code of ethics

Personnel within an organization have a responsibility to understand and support their organization's code of ethics. These vary between organizations, based on the objectives and mission of the organization, but you can often find a reference to the organization's values and/or a specific code of ethics in its security policy.

True or false? An organization's ethics program should include support of all applicable regulatory requirements.

Answer: *True.* At the very least, an ethics program will remind employees of rules and regulations that apply to the organization and the organization's commitment to abide by these rules and regulations.

Many organizations have an ethical statement that is developed and/or approved by the executive leadership. An ethical statement will often include guiding principles that personnel can use to help them make decisions. These are especially useful when personnel are faced with an ethical dilemma. When faced with a situation that presents two seemingly unacceptable outcomes, personnel can make their decision based on the guiding principles.

True or false? An administrator who comes across objectionable material on a user's computer should immediately report it to the police.

Answer: *False.* Personnel faced with an ethical problem that is not breaking the law should report the problem internally.

Objectionable material often isn't illegal. However, someone within the organization should make the determination about whether it is illegal or not and what action to take.

> **EXAM TIP** When faced with an ethics issue within an organization, the first choice for help and support in resolving the issue should be from personnel in the organization. That is, personnel have an obligation to support their organization's code of ethics. An exception is when a law demands a different action. For example, if individuals are threatening to harm themselves or others, law enforcement personnel should be called.

> **MORE INFO** The Internet Architecture Board (IAB) published RFC 1087, "Ethics and the Internet," which can be accessed here: *http://tools.ietf.org/html/rfc1087*. Even though this is an older RFC, it is still recognized as a standard for online ethics issues.

Can you answer these questions?

You can find the answers to these questions at the end of this chapter.

1. What are the mandatory elements of the Code of Ethics that all (ISC)2 candidates must abide by?
2. When faced with an ethical issue within an organization, who should be consulted?

Objective 9.3: Understand and support investigations

Investigations are performed in response to security incidents to identify what happened, what allowed it to happen, and what can be done to prevent it from happening again. Additionally, investigations collect evidence that can be used in a court of law to prosecute criminals when appropriate. At the core of most legal investigations is Locard's exchange principle. This principle states that a criminal will leave something behind at the crime scene and take something from it. Investigators assume this to be true and take the time to identify both what the criminal took and what the criminal left behind. This same concept applies to computer crime investigations.

Exam need to know...

- Policy, roles, and responsibilities (e.g., rules of engagement, authorization, scope)
  For example: When is it acceptable to launch a counterattack? What is a virtual incident response team?
- Incident handling and response
  For example: What should an incident response team do before attempting to isolate a system? When isolating a system, what is a concern beyond containment?
- Evidence collection and handling (e.g., chain of custody, interviewing)
  For example: What can corrupt evidence when it is collected? What is used to verify that the evidence has been protected?
- Reporting and documenting
  For example: What is the purpose of a final report? Who should have access to the final report?

Policy, roles, and responsibilities (e.g., rules of engagement, authorization, scope)

When security incidents occur, it's important for personnel to understand their responsibilities and how they can support security incident investigations. These responsibilities are outlined in security policies, with more details included in incident response plans. Many organizations have dedicated incident response teams that respond after a security incident has been reported and begin an investigation.

True or false? An incident response plan will outline the specific responsibilities of members of an incident response team.

Answer: *True*. Organizations that have incident response teams will also have an incident response plan to specify roles and responsibilities of personnel on the team.

An incident response team can be a dedicated team with most team members performing incident response tasks as their primary job responsibility. However, this is cost-prohibitive for most organizations. Instead, many organizations have a virtual team that comes together only when needed. Each team member will have a primary job role elsewhere in the organization, but when an incident occurs, the team members come together to respond to the incident.

True or false? Personnel on a virtual incident response team do not need any specialized training other than the training they have for their primary job.

Answer: *False*. Incident response team members require specialized training to ensure that they know their responsibilities when responding to an incident and the best methods to respond.

Effective incident response requires a significant amount of specialized knowledge not normally available in other job roles. For example, many people do not understand how to protect, collect, handle, and investigate forensic evidence. Mistakes in this process can invalidate evidence and allow a criminal to escape prosecution for a crime.

> **EXAM TIP** Members of an incident response team require specific training to fulfill their job roles. A virtual incident response team is composed of employees, contractors, and/or consultants that have other primary jobs but come together to respond to an incident when required.

> **MORE INFO** The SANS Institute Reading Room has a concise overview of a computer incident response team, including roles and responsibilities. You can access it here: *http://www.sans.org/reading_room/whitepapers/incident/computer-incident-response-team_641.*

True or false? Counterattacks by incident response teams are authorized as long as the source IP address has been verified as a valid IP address used by the attacker.

Answer: *False*. Counterattacks are rarely authorized by members of an incident response team. Instead, the incident response team focuses on verifying that an incident occurred, taking steps to minimize the damage, and collecting evidence that can be used in an investigation.

Organizations typically have clear guidelines on rules of engagement for counterattacks. Unless the organization specializes in seeking and attacking criminals, the rules of engagement almost always dictate that attackers should not be counterattacked. First, most attacks against other computers are illegal and cannot be justified by saying "he attacked first." Some law enforcement and military agencies have teams with specialized training to launch attacks and have clear rules of engagement defining when counterattacks are authorized, but they are the exception.

It's also important to realize that it's difficult to definitively identify the attacker. IP addresses are usually spoofed, so the source IP address in a packet used in an attack can't be trusted as proof of the attacker. Additionally, attackers often use innocent user systems as proxies to launch attacks. An infected system can be directed to launch an attack, and a counterattack will hit this innocent user's system.

> **EXAM TIP** Personnel should never launch a counterattack without specific authorization to do so, which is normally defined in clear rules of engagement. A counterattack might violate laws and will often result in attacking innocent individuals instead of the criminal who launched the original attack.

Incident handling and response

One of the goals of security is to prevent incidents. However, they occur, and when they are detected, effective incident handling and response procedures help limit the damage to the organization. From an investigation perspective, one of the key goals during the response is to ensure that evidence is protected.

True or false? One of the first steps that an incident response team must take after an incident is to verify that a security incident has occurred.

Answer: *True*. The first step is to verify that a security incident or computer crime has occurred.

False alarms and false positives occur all the time, but they are not actual incidents. Before going too far into the response, it's important to verify that an attack, incident, or crime either has taken place or is currently occurring.

True or false? After verifying that a system has been compromised from a security attack, the first thing to do is to isolate or contain the affected system.

Answer: *True*. Isolating a system contains the attack. This can often be done by removing a cable from a network interface card (NIC) to prevent any communication from the compromised system.

Turning off power to a system will isolate the system, but this is not recommended because it also destroys key forensic evidence. The goal is to isolate the system while also protecting the evidence. Disconnecting the network cable meets both objectives.

> **EXAM TIP** Incident response procedures should start by verifying that an incident is occurring or has occurred. Next, a containment stage isolates or contains the system to limit the damage or spread of the incident. To ensure that evidence remains available for collection, the affected system should not be powered down. Similarly, manipulating files or anything on the computer is not recommended until evidence has been collected, because any activity can potentially modify the evidence.

> **MORE INFO** Chapter 7, "Operations security," includes more details about incident response in the context of Objective 7.3. The steps involved in an incident response are detection, response, reporting, recovery, and remediation/review.

Evidence collection and handling (e.g., chain of custody, interviewing)

If evidence must be collected after an incident, several steps must be followed to protect this data. The goal is to ensure that the evidence is not modified and can be used in a court of law.

True or false? A chain of custody log provides assurances that data has not been modified.

Answer: *False*. The chain of custody log provides assurances that data (or any type of evidence) has been controlled after it was collected.

A chain of custody log documents the location of evidence and who controlled it at any time since it was collected. Controlled evidence is not susceptible to tampering. Other methods are used to ensure that the evidence is not modified when it is collected or during the analysis stage of the investigation.

> **EXAM TIP** Forensic collection procedures ensure that data is not modified when it is collected. After it is collected, a chain of custody log documents where evidence has been at any given time and provides assurances that it has been controlled since it was collected. Forensic analysis procedures ensure that the data is not modified during the analysis steps.

True or false? Effective access control practices can provide accountability in a forensic investigation.

Answer: *True*. If users are accurately identified and securely authenticated, their actions can be recorded in security logs, providing accountability in forensic investigations.

Security logs can be used to prove that an individual took specific actions. A forensic investigation uses an audit trail to re-create events such as when a user logged on, what the user did while logged on, and when the user logged off. If the user was involved in malicious or unacceptable activities, the results of the investigation can be used to hold the individual accountable.

True or false? Admissible evidence in a court of law must be relevant and reliable.

Answer: *True*. Relevant evidence has a relationship to the findings and can either prove or disprove a point. Reliable evidence is data that has been obtained from an original source, has not been modified, and has been protected since it was collected.

True or false? An interrogation is an interview used to gather information about a crime.

Answer: *True*. If information from an interview is intended to be used in a court of law, it is considered an interrogation. Personnel conducting interviews and interrogations should be trained and understand the legal implications.

> **EXAM TIP** A thorough understanding of relevant laws related to investigations is critical if evidence is to be used. If evidence is not collected and handled properly or if an individual's rights are violated during an interview or interrogation, a criminal case can be dismissed.

Reporting and documenting

After an incident and investigation, it's important to document the results in a report. The report will include information about the incident, what allowed it to happen, details of the investigation, and summary conclusions. In many cases, the report will include recommendations for additional actions.

True or false? An incident report will include a determination about whether the source of the incident was internal or external.

Answer: *True*. After the incident has been verified and isolated and data has been collected and analyzed, investigators attempt to determine the source of the incident.

During the investigation stage, investigators attempt to discover the root cause of the incident. They attempt to identify what allowed the incident to occur and what can prevent it from recurring. If an internal source caused the incident, investigators normally have free access to documentation to investigate the incident. If an external source caused the incident, they often need subpoenas or other legal documents to gather the information that they need. Many of the details of these steps will be included in the report.

True or false? Copies of the full investigation reports should be shared with personnel throughout the organization.

Answer: *False*. Investigation reports often include sensitive information and should have limited distribution to only executive management and security personnel.

In some cases, different versions of investigation reports are created for different audiences. For example, a limited version of the report might be created outlining the incident, its impact on the organization, the cause, and steps taken to correct it. This can be shared with all personnel in a security training event to help them understand the reasoning behind certain security requirements.

> **EXAM TIP** Documentation and reporting of investigation results is valuable to the organization to prevent a repeat of the same incident and also to help with planning. However, these reports should be protected. Often, reports will identify response steps, security procedures, current vulnerabilities, and other information that can be valuable to an attacker. Only executive management and security personnel should have access to the full report.

Can you answer these questions?

You can find the answers to these questions at the end of this chapter.

1. A member of an incident response team has located evidence about the identity of an attacker and knows he can successfully launch an attack to disable the attacker's computer equipment. When is this acceptable?
2. What should an incident response team member do when first learning about a potential incident?
3. What is a primary concern related to evidence after it is collected, and how is this concern addressed during an investigation?
4. Who should have access to the full security incident investigation report?

Objective 9.4: Understand forensic procedures

Computer forensics techniques use specialized tools, along with scientific knowledge, to capture and analyze evidence. When done correctly, the forensic evidence can be used in a court of law to prosecute criminals with assurances that the evidence is the same evidence that was originally collected and was not modified during collection or analysis. When forensic procedures are performed incorrectly, valid data can be declared inadmissible, allowing criminals to go free. Worse, if data is not captured, analyzed, and protected correctly, innocent people can be found guilty of crimes they didn't commit.

Exam need to know...

- Media analysis
  For example: What should be avoided when copying data that will be used for forensic analysis? What is used to copy a disk drive?
- Network analysis
  For example: What needs to be configured on a switch when capturing network traffic going through the switch? What are some elements in a packet worth analyzing?
- Software analysis
  For example: How can a botnet command and control center be located? What are some capabilities of a forensic toolkit?
- Hardware/embedded device analysis
  For example: Is it possible to retrieve data from a smartphone or tablet device?

Media analysis

Forensic procedures used with media can capture and analyze data from media such as disk drives, optical disks, USB flash drives, and backup tapes. When data is captured, a primary consideration is ensuring that the data is not modified.

True or false? Before analyzing data on a disk drive, a forensic analyst should create a bit-level image of the drive.

Answer: *True.* A bit-level image is an accurate reproduction of the disk that does not modify the data.

The process of analyzing data can modify the data and make it inadmissible as evidence. Instead of analyzing the original disk, forensic experts create a bit-level image and then analyze the bit-level image. Later, it's possible to create another bit-level image and reproduce the analysis to achieve the same results.

Regular copying tools within the operating system often modify the data and metadata during the copying process. If the two disks are different sizes, an operating system copy tool might even change the size of the copied file and modify information in a file's slack space. In contrast, a bit-level program copies everything bit by bit.

NOTE Slack space in a file is the extra space between the end of the data in a file and the end of the cluster used for a file. For example, a file might be stored in several 2,048-byte clusters, but the last cluster might hold only 40 bytes of data. The remaining 2,008 bytes in the last cluster is slack space. Slack space often holds data from a previously deleted file or, in some cases, data that was in memory when the file was saved. Forensic tools can recover data in slack space. However, if improper collection procedures are used, the original information in the slack space is lost.

True or false? A write blocker is a forensic tool used to prevent any write commands from being sent to a disk.

Answer: *True*. Write blockers either have a white list of specific read commands that are allowed and they block all others or have a blocked list of all write commands that are not allowed.

Write blockers are typically hardware devices, although there are some software-based write blockers. Hardware-based write blockers are available that can read from just about any disk drive interface, including USB and other portable media devices.

In some cases, the goal is to return a system to operation as quickly as possible. If evidence needs to be collected, forensic tools can capture the evidence without modifying it and then return the system to operation.

EXAM TIP One of the important concepts to remember when collecting evidence is that the evidence should not be modified at any time. It shouldn't be modified during collection or analysis and should be protected from modification while it is held. To preserve the integrity of the original evidence, data captured for forensic analysis should be captured with bit-level copy programs.

True or false? Combining data from multiple sources, such as copying it onto a single medium, can corrupt forensic evidence.

Answer: *True*. Forensically sound data should not be combined with other data samples.

Copying data from different sources onto the same USB or optical disc can inadvertently modify the data. If the data is intended to be used as forensic evidence, this action can result in the evidence being deemed inadmissible. This combination can modify metadata, such as some of the attributes of the files and the slack space.

EXAM TIP Evidence on a computer should be captured before the computer is turned off or rebooted. Additionally, personnel should not manipulate any files or programs on the computer until evidence is captured. In some cases, attackers leave behind programs designed to erase their tracks. These programs can be triggered by certain user activity.

True or false? Time stamps used on files are valuable as evidence during an investigation.

Answer: *True*. Time stamps include information such as when the file was created, saved, accessed, or archived.

> **EXAM TIP** Many time stamps can be modified during the collection of the data. Using a bit-level program to capture one or more copies of the disk will not modify any time stamps. Investigators analyze the copies and preserve the original evidence, including the original time stamps.

True or false? Hashing techniques can be used to verify that data has not been modified during the capture.

Answer: *True*. By comparing the hash of data before it is captured and after it is captured, it's possible to verify that the data has not been modified if the hash is the same.

Hashing is commonly used to verify the integrity of files or messages, and it can also be used with evidence. A hashing algorithm will always produce the same hash when executed against the same data. As long as the resultant hash is the same, it verifies that the original data is the same.

> **MORE INFO** Hashing is covered in more detail in Chapter 5, "Cryptography," in the context of Objective 5.3. Overall, hashing helps preserve integrity, and when used with forensics, it helps preserve the integrity of evidence.

Network analysis

Network analysis captures traffic sent over the network. A packet analyzer (also called a sniffer) can be connected to a router or switch and capture all traffic sent through the device.

True or false? Port mirroring is used with a sniffer to capture all traffic sent through a switch.

Answer: *True*. A port mirror sends a copy of all traffic sent through a network device to a specific port. The sniffer is connected to this port and can capture all the data for analysis.

Data can also be captured from available logs. This includes firewall logs, intrusion detection/prevention system logs, and any other log that captures network traffic.

True or false? Packet digest information can be captured for analysis instead of entire packets.

Answer: *True*. To limit the amount of captured data, a software program can save just some of the information from captured packets.

Information saved in a packet digest can often be modified based on the desire of the analyst. For example, the analyst might choose to save the payload along with the source and destination IP addresses, media access control (MAC) addresses, ports, or any combination of this data.

True or false? A packet analysis program can reliably identify the IP address of a computer involved in an attack.

Answer: *False*. While packets include source IP address information, attackers often spoof the source IP address, so this information isn't reliable for identifying the IP address of the attacking computer.

> **MORE INFO** There are many tools that can be used to collect IP traffic. Microsoft's Network Monitor is available for free from their download site (*http://www.microsoft.com/download/*) by searching on Network Monitor. A newer version, called Microsoft Message Analyzer, is discussed on the following page: *http://blogs.technet.com/messageanalyzer*. Wireshark is a popular tool available for free here: *http://www.wireshark.org/*. Cisco has developed NetFlow, a protocol used to collect IP traffic information, which is discussed on this page: *http://www.cisco.com/en/US/docs/ios/11_2/feature/guide/netflow.html*.

With the use of other available data, packet analysis programs are useful in identifying in-house computers involved in an attack. For example, the MAC and IP address of the source computer can be compared with the Dynamic Host Configuration Protocol (DHCP) logs. These logs will show when a specific IP address was assigned to a specific MAC address. It's also possible to look at logs on individual computers to identify when an IP address was assigned, and at firewall or router logs to identify traffic passing through the devices. Put simply, the single packet isn't definitive proof of malicious activity. However, when combined with other available evidence, it can help investigators validate the information.

Software analysis

Software analysis has two meanings in the context of computer forensics. First, when experts analyze software code looking for malicious content, they are performing software analysis of this code. Second, there are many different software tools available to assist experts in forensic investigations. They use these software tools to assist in their analysis of available data.

True or false? A malicious software (malware) expert can use software analysis techniques to identify the command and control server used in a botnet.

Answer: *True*. Software analysis tools are used to analyze various types of malware, and they can identify information about botnet servers.

Botnets are controlled by bot herders through one or more command and control servers. Infected computers (commonly called zombies) periodically contact the command and control server to download instructions. By analyzing the code in the infected computer, experts can gather details about the command and control servers.

It's sometimes possible to use network logs to identify the command and control servers. For example, if hundreds of internal computers are joined to the same botnet, they might be contacting the command and control server at the same time. This isn't as apparent if only one or two systems are joined to the botnet because the contacts look just like any other Internet connection.

Malware developers often include bogus information for false command and control servers. After analyzing the code, the forensic expert might have a list of hundreds of possible command and control servers, with only one actually being used. However, it is now possible to match this list against network analysis traffic to determine which servers are valid.

True or false? A forensic toolkit can capture data from volatile memory and save it for analysis.

Answer: *True*. Forensic toolkits can capture data from volatile memory. Copies of this data can then be analyzed without affecting the original information.

In addition to performing bit-level copies, forensic toolkits can also capture data within physical and virtual memory. This includes any data from running processes, which often reveal recent activity that can be valuable during an investigation. Data in physical memory is volatile and will be lost when the computer is turned off. Data in virtual memory (often called a swap file) is within a file stored on the disk drive. A logical shutdown of the computer often erases data in the swap file, but it will remain intact if the computer is abruptly turned off. The swap file will be overwritten when the computer is restarted.

> **MORE INFO** Forensic experts have a variety of different tools that they can use to create bit-level copies and to perform other analyses. Common tools include Sysinternals, EnCase, and the Forensic Toolkit. Some links to information about these tools are as follows: http://computer-forensics.sans.org/blog/tags/sysinternals, http://www.guidancesoftware.com/encase-forensic.htm, and http://computer-forensics.sans.org/community/downloads.

Software toolkits include multiple features. The following is a list of some of the features available in EnCase, sold by Guidance Software, but similar features can be found in most forensic toolkits:

- Tools can capture data from almost any source, including memory, disk, files, web histories, chat sessions, and more.
- Forensically sound procedures acceptable for evidence are used to ensure that exact duplicates of data are captured without modifying the original data.
- File recovery capabilities include the ability to identify recently deleted files and recover deleted files and partitions.
- Automated capabilities help eliminate known good files from the list of files that need to be analyzed. This reduces the overall analysis time.
- Reporting capabilities include both summary and detailed reports. Reports provide actionable data that can be used by management and/or in a court of law.

Hardware/embedded device analysis

Forensic tools are also available to analyze different hardware devices, including smartphones, tablets, and other computing devices embedded within systems, such as within a car.

MORE INFO This section is focused on portable and embedded devices, but it's important to realize that forensic information can be retrieved from workstations and servers also. The "Media analysis" section mentioned methods used to retrieve information from disks, and the "Software analysis" section mentioned methods used to retrieve information from volatile memory.

True or false? Forensic analysts can retrieve information from smartphones or tablet devices that might indicate where the device has been at specific dates and times.

Answer: *True*. Data stored on these devices often includes location data gathered from global positioning systems (GPSs). When this data is available, forensic experts and others can retrieve it.

The specific details of how to retrieve forensic information from devices is often dependent on the hardware. However, there are many training courses that provide information about how to retrieve information from various devices.

EXAM TIP If data is stored on any type of device, it is very likely that there is a method of retrieving and analyzing this information. The method might be well known and located through a quick Bing search on the Internet, or it might be known by only a select group of people. Forensic experts are usually in the group that knows how to retrieve evidence from any device and should be consulted if the information is needed for an investigation.

MORE INFO The article titled "Got an iPhone or 3G iPad? Apple is recording your moves" explains how location data has been stored on Apple devices. Many training events have provided details to forensic analysts about how to retrieve this information as part of a forensic investigation. You can read the full article here: *http://radar.oreilly.com/2011/04/apple-location-tracking.html*.

Can you answer these questions?

You can find the answers to these questions at the end of this chapter.

1. What type of tool should be used when creating a copy of a disk for a forensic investigation?
2. What type of data can be captured and analyzed with a protocol analyzer?
3. What do forensic analysts use to capture and analyze data that will be used as evidence?
4. Smartphones have the capability to store data, but can any of this data be retrieved for an investigation?

Objective 9.5: Understand compliance requirements and procedures

Organizations that are covered under specific regulations must ensure that they follow all the relevant compliance requirements and procedures. The specific laws that apply to the organization, based on its function and location, make up the regulatory environment. Often these regulations require the organization to perform regular audits and report the results to different entities. These requirements often take a significant amount of time and effort, but the penalties of noncompliance are worse than the cost to comply.

Exam need to know...

- Regulatory environment
  For example: What are the different types of laws? What is an example of a nonregulatory requirement?
- Audits
  For example: Can audits be performed internally? What is the purpose of a financial audit?
- Reporting
  For example: Who is responsible for submitting reports?

Regulatory environment

The regulatory environment is composed of various laws and regulations that apply to an organization. Depending on the location of the organization and the type of business, the regulatory environment can vary.

Laws are divided into regulatory, criminal, and civil categories. Regulatory laws are the primary concern when evaluating the regulatory environment, but it's important to understand the following differences among the types of laws:

- Regulatory laws are created by government entities and define specific standards that organizations must follow. Failure to follow these regulations typically results in fines or other penalties against the organization.
- Criminal laws are used to protect the public from criminal actions. A government entity prosecutes individuals that break these laws. When found guilty, individuals can be punished with fines and/or jail time.
- Civil laws (sometimes called tort laws) are used to provide a remedy for entities that have been wronged. Wronged entities (both individuals and organizations) can use civil lawsuits to recover damages and other losses.

True or false? A U.S. financial organization that keeps employee data on selected health plans is required to comply with the United States Health Insurance Portability and Accountability Act (HIPAA).

Answer: *True*. Organizations that handle or maintain protected health information (PHI) are identified as covered entities and are required to comply with HIPAA.

HIPAA has specific requirements for the protection of health information, and it doesn't apply only to medical providers. PHI includes information that can be used to identify the individual as well as the following types of information:

- Information created or received by a health care provider, health plan, employer, or health care clearinghouse
- Information related to the past, present, or future physical or mental health or condition of an individual
- Information about the provision of health care to an individual or about the past, present, or future payment for the provision of health care to an individual

EXAM TIP Many people think that HIPAA applies only to medical providers. However, it applies to many organizations that aren't involved in health care. Any organization that provides a health care plan and collects or maintains PHI is identified as a covered entity and must comply with HIPAA.

MORE INFO National Institute of Standards and Technology (NIST) Special Publication (SP) 800-66, "An Introductory Resource Guide for Implementing the Health Insurance Portability and Accountability Act (HIPAA) Security Rule," describes key security management processes used to comply with the HIPAA standard. You can download it here: *http://csrc.nist.gov/publications/PubsSPs.html*. HIPAA is also mentioned in Chapter 3 in the context of third-party governance in Objective 3.6.

True or false? The Payment Card Industry Data Security Standard (PCI DSS) is an example of a self-regulatory regulation.

Answer: *True*. The requirements under PCI DSS are not law but instead are standards that are self-regulated by the merchants and the PCI Security Standards Council. In contrast, many regulations are laws enacted by a government.

MORE INFO Chapter 6, "Security architecture & design," explores PCI DSS more in the context of Objective 6.2. PCI DSS includes 12 specific requirements that merchants agree to abide by. Most merchants validate that they are complying with the requirements by completing a self-assessment questionnaire.

Audits

Many laws and regulations require an organization to regularly complete audits. Audits evaluate an element of an organization to verify that it is in compliance with governing requirements. Financial audits examine the financial statements of an organization and provide an opinion on their accuracy.

Other types of audits check processes or procedures to verify that an organization is following established rules and guidelines. For example, an organization can perform an internal account access audit to verify that the principle of least privilege is being followed. Similarly, regulations can require a company to complete audits to verify compliance with specific laws.

True or false? The Sarbanes-Oxley Act of 2002 requires principal officers in the company to personally verify the accuracy of audit reports and financial statements.

Answer: *True*. The Sarbanes-Oxley Act of 2002 (SOX) applies to all publicly traded companies in the United States and requires officers such as CEOs and CFOs to personally verify the accuracy of reported financial statements.

Sarbanes-Oxley also provided standards for external auditor independence. It introduced auditor reporting requirements, audit partner rotation, and prevents auditors from providing consulting or other non-audit related services to the same company. Combined, these standards and requirements help limit conflicts of interest by the auditors.

> **MORE INFO** Chapter 3 includes additional information on Sarbanes-Oxley in the context of Objective 3.6.

Reporting

Some regulations require organizations to report their compliance in one way or another. The method and frequency of reporting varies depending on the regulation.

True or false? Organizations that are covered by PCI DSS must report compliance annually.

Answer: *True*. PCI DSS requires merchants to validate their compliance when they first subscribe to PCI DSS and annually afterwards.

Smaller companies need to complete only a self-assessment questionnaire to report compliance. Larger organizations exceeding specific thresholds of credit card transactions must contract with a Qualified Security Assessor to complete an audit and report compliance of the organization.

True or false? If the organization contracts a third party to complete an audit, the organization is not responsible for the report.

Answer: *False*. If a regulation requires a report from an organization, the organization cannot delegate that responsibility. Organizations can contract an outside entity to complete the audit, but they still have a responsibility to ensure that reporting requirements are met.

Can you answer these questions?

You can find the answers to these questions at the end of this chapter.

1. What are three common types of laws?
2. What is the purpose of a financial audit?
3. How often are reports required for compliance under a law or regulation?

Objective 9.6: Ensure security in contractual agreements and procurement processes (e.g., cloud computing, outsourcing, vendor governance)

Security also needs to be considered when organizations enter into contractual agreements with other organizations. Protecting against losses of confidentiality, integrity, or availability remains a concern, even when the work or processes are outsourced. Cloud computing is becoming more and more common and making it easier for organizations to rent or lease cloud computing services. However, the decision to use these outside services should not be taken without considering potential risks.

Exam need to know...

- Contractual agreements and procurement processes
  For example: What are the different types of laws? What is an example of a nonregulatory requirement? What is provided in a contract? Does the cloud-based service back up the data?

Contractual agreements and procurement processes

Organizations commonly use contractual agreements and procurement processes when creating and marketing their goods and services. The goal is to ensure that they receive what they expect while also minimizing their risks.

True or false? Contracts are not used when dealing with vendors.

Answer: *False*. Organizations often contract with vendors to ensure that they receive consistent pricing and services.

Contracts are agreements between two or more entities, and they create a legal obligation between the different parties. They can be used for any entity with which an organization wants to do business. This includes personnel hired on a contract basis, external companies that promise to provide contract personnel, and agreements to provide products or a specific level of service.

> **EXAM TIP** Contracts are legal documents and need to be reviewed by appropriate personnel. It's common for legal advisors to slip in terminology that will greatly benefit one party in a contract agreement but not the other party. When all parties seek legal advice, it helps ensure that the contract is fair for all or at least that all parties are equally aware of the provisions.

Cloud computing is loosely defined as computing that includes the use of applications, infrastructure, and/or data over a network. Typically the network is the Internet, but it's also possible to create private clouds within an organization that can use private communication links or semi-private virtual private network (VPN) communication links.

True or false? All organizations that sell data storage via the cloud ensure that data is adequately protected and regularly backed up.

Answer: *False.* All cloud services do not provide the same level of data protection or backups.

When looking for a company providing cloud services, it's important for a company to exercise due diligence to ensure that the cloud provider employs adequate security. Some providers have extensive security programs to ensure that the data is protected. They also perform consistent backups and regularly check the backup process to ensure that data is always available according to contract terms. Other companies aren't as secure and provide very few guarantees in their contractual agreements.

True or false? Some companies provide cloud computing services allowing users to run software from thin clients via the Internet.

Answer: *True.* The software often runs through the web browser, and almost any device that can run a web browser can run the application.

Cloud computing is often divided into the following categories:

- **Software as a Service (SaaS)** Users can run software applications from a web browser. For example, Microsoft hosts Office 365, which is the full suite of Microsoft Office applications that users can access via the Internet through Internet Explorer. Documents used through this service can also be viewed and modified with Microsoft Office desktop applications.
- **Platform as a Service (PaaS)** Users have access to a computing platform via the Internet, such as an operating system or an application development engine. This is sometimes used by developers during application development and will usually be running on a virtual machine.
- **Infrastructure as a Service (IaaS)** This includes servers (typically running as virtual machines) and other supporting infrastructure, such as network devices and storage space. This allows an organization to easily scale up operations to meet short-term demand. For example, an e-commerce site can rent IaaS services to meet the demand of a holiday buying season without the cost of procurement and long-term maintenance.

MORE INFO You can read more about Office 365 here: *http://www.microsoft.com/office365*. Also, Microsoft maintains an extensive site on private cloud solutions that an organization can create for its employees at *http://technet.microsoft.com/cloud/private-cloud*.

Key concerns that many organizations have when contracting with a cloud computing company include the following:

- **Confidentiality** Organizations can lose control of the data, resulting in a loss of confidentiality.
- **Integrity** When another organization controls the data, it is susceptible to unauthorized modification.

- **Availability** Systems and data controlled by a third party are susceptible to outages.

> **EXAM TIP** All cloud-computing services are not the same. When using or contracting for cloud-based services, it's important to verify that the outside company provides an adequate level of protection. A primary consideration is ensuring that the company can meet the confidentiality, integrity, and availability requirements of the organization using the service.

True or false? Organizations that use external companies to handle data covered under the Gramm-Leach-Bliley Act (GLBA) must enter into a contract with the company to ensure that data is protected.

Answer: *True*. Companies that outsource some of their work must ensure that the external companies protect their data.

Both GLBA and HIPAA require covered businesses to enter into contracts to ensure that data is protected under the governing law.

> **MORE INFO** GLBA was passed in 1999, and it requires financial institutions to provide privacy notices to customers and give customers the option of preventing their information from being shared with others. You can download a summary of the GLBA here: http://www.symtrex.com/pdfdocs/glb_paper.pdf.

Can you answer these questions?

You can find the answers to these questions at the end of this chapter.

1. What is the purpose of a contract?
2. What should be considered when using a cloud-based service?
3. Which type of cloud-based service allows users to run software applications via a web browser?

Answers

This section contains the answers to the "Can you answer these questions?" sections in this chapter.

Objective 9.1: Understand legal issues that pertain to information security internationally

1. Click-fraud is a computer-assisted computer crime. Criminals use software to simulate clicks on advertisements, and they receive funds for each simulated click.
2. Trial software allows individuals to use software for a specific time period (such as 30 days). This software licensing category is sometimes called shareware.
3. A terms of use contract used on websites is also known as a browsewrap contract.

4. The OECD provides guidelines used to simplify trans-border data flow transactions.
5. Identity theft is a crime where an individual's PII is stolen and used by the criminal for personal gain. Organizations have a responsibility to protect PII, which can reduce the incidence of identity theft.

Objective 9.2: Understand professional ethics

1. The mandatory elements of the Code of Ethics are the preamble and the four canons. The four canons, in the proper order, are as follows:
   a. Protect society, the commonwealth, and the infrastructure.
   b. Act honorably, honestly, justly, responsibly, and legally.
   c. Provide diligent and competent service to principals.
   d. Advance and protect the profession.
2. Someone within the organization should be consulted to resolve ethical issues. This is often an employee's supervisor. However, depending on the situation, someone else might need to be consulted, such as legal personnel or experts within the organization's human resources department.

Objective 9.3: Understand and support investigations

1. It is rarely acceptable to launch a counterattack against an attacker. Personnel should do this only when it is authorized under their rules of engagement and/or when they are directed to do so by senior personnel.
2. When first learning about an incident, incident team members must investigate the situation to verify that it is an actual incident. After validating a security incident occurred, the next step is to contain the incident by isolating affected systems.
3. Protection of evidence is a primary concern after it is collected. A chain of custody log is used to verify that the evidence has been protected since it was collected. During collection and during analysis, a primary concern is ensuring that the evidence is not modified.
4. Only security personnel and senior management should have access to full security incident investigation reports. These reports will often have sensitive information, and they should be protected. Limited versions are sometimes created to summarize the data for specific audiences.

Objective 9.4: Understand forensic procedures

1. Bit-level copy tools should be used when creating a copy of a disk. These tools will copy all the data exactly as it is on the original disk without modifying it.
2. Any data included in a network packet or frame can be captured by a protocol analyzer (also called a sniffer). This includes IP addresses, MAC addresses, ports, and payload data. An analyst can use this to determine the destination but should be suspicious of the source address because it can be spoofed.

3. Forensic analysts use specialized toolkits that include features to retrieve data from just about any source while ensuring that the data is not modified during capture. These tools also have advanced capabilities to streamline the analysis process. EnCase is a toolkit used by many forensic experts, but other toolkits are available.
4. Data can be retrieved from almost any device that stores data, including smartphones. The only challenge is having the right tools and the knowledge to do so, and forensic experts usually have both. Unfortunately, many criminals have access to the same tools and knowledge.

Objective 9.5: Understand compliance requirements and procedures

1. Three common types of laws are regulatory (which are often the most relevant to organizations for compliance purposes), criminal, and civil (also called tort laws).
2. A financial audit attempts to validate the financial statements of an organization.
3. Individual laws and regulations have various reporting requirements. Annual reporting is common, but it is not necessarily the requirement for any regulation.

Objective 9.6: Ensure security in contractual agreements and procurement processes (e.g., cloud computing, outsourcing, vendor governance)

1. A contract is used to provide a legal obligation between two or more entities. It is a formal agreement that binds each party to comply with their agreement.
2. Organizations should verify that a cloud-based service can meet its confidentiality, integrity, and availability requirements.
3. SaaS is a cloud-based service that allows users to run software applications through a web browser. For example, Office 365 allows users to run the Microsoft Office suite through a web browser.

CHAPTER 10

Physical (environmental) security

The Physical (Environmental) Security domain covers threats and vulnerabilities related to the physical environment and countermeasures an organization can take to reduce the associated risks. The goal is to protect all of an organization's resources, including the people, the facilities, the equipment, and physical media. Indirectly, this helps protect data stored on any of the equipment or media. A primary goal of physical security is to prevent unauthorized personnel from accessing a facility or its assets.

This chapter covers the following objectives:

- Objective 10.1: Understand site and facility design considerations
- Objective 10.2: Support the implementation and operation of perimeter security (e.g., physical access control and monitoring, audit trails/access logs)
- Objective 10.3: Support the implementation and operation of internal security (e.g., escort requirements/visitor control, keys and locks)
- Objective 10.4: Support the implementation and operation of facilities security (e.g., technology convergence)
- Objective 10.5: Support the protection and securing of equipment
- Objective 10.6: Understand personnel privacy and safety (e.g., duress, travel, monitoring)

Objective 10.1: Understand site and facility design considerations

A primary element of physical security is starting with a secure site. Design measures ensure that any access by unauthorized personnel is deterred, denied, detected, and/or delayed. When detected, a response attempts to stop the unauthorized access. A core topic in the site and facility design consideration is crime prevention through environmental design (CPTED) strategies.

Exam need to know...

- Site and facility design considerations
  For example: What are three primary strategies of CPTED? How many physical barriers should a facility include?

Site and facility design considerations

Just as with any type of security, physical security should be considered during the design stage of a facility. If an organization is building or redesigning a facility, there are many physical security elements that it can include in the design to aid physical security.

When evaluating a site as a possible business location, several factors that might impact the viability of the business should be evaluated. This includes any type of natural disaster risks such as flooding, hurricanes, or earthquakes. The crime rate in the local area and the distance from police and fire stations should also be evaluated.

True or false? An example of a natural access control is the inclusion of a staircase in the middle of a two-story building using open rails instead of walls.

Answer: *False*. This is an example of natural surveillance, one of the three primary strategies of CPTED, not a natural access control strategy.

The following three CPTED strategies are commonly used to reduce crime:

- **Natural surveillance** The basic goal of this is to increase the threat that an attacker can be seen. Facilities are designed to maximize visibility of people throughout the building. For example, stairs and halls are designed with an open architecture so that people are easily viewable. This reduces the paths of escape for unauthorized personnel. Similarly, outside lighting is designed so that all areas are lit.

- **Natural access control** This uses clearly defined entries and exits. Often there will be a single exit for public personnel with design elements that guide visitors directly to reception areas. In some cases, there might be multiple controlled entries for employees separate from the public entry. Within the building, restricted spaces such as a server room or data center will normally have a single entry/exit point to restrict and control access.

- **Natural territorial reinforcement** Landscaping is used to show that the facility is well maintained and occupied. Fences, signs, and pavement help to define the boundaries and clearly indicate to public personnel when they cross into an unauthorized area. This keeps out the general public and also helps unauthorized personnel stand out when they do enter.

Windows' placement and strength are commonly considered in the design of a building. Ideally, they should be difficult to access, and when they are accessible, they should be strong to resist threats. Some glass has embedded wires to resist access.

EXAM TIP The strongest type of window glass is glass-clad polycarbonate, which comes in different bulletproof levels. Acrylic glass is very strong, but it is also combustible and emits toxic fumes when it burns, so its use is sometimes restricted by fire codes.

True or false? A single strong barrier is the best protection against physical attacks.

Answer: *False*. Most facilities use a layered approach with multiple barriers. Even if an attacker breaches one layer of security, the attacker will still have multiple additional layers to bypass.

It's common for an organization to have at least three layers of security. The outer layer could be a fence around the property or the building. Access to the building is restricted to certain entries for visitors, providing another boundary. Access to the internal work spaces is controlled by guards, cipher-locked doors, or man traps, which function as inner boundaries with more restrictive controls. These additional controls ensure that only employees and authorized visitors are granted access. Restricted areas within the internal work spaces (such as a data center or server room) are further restricted (such as by using cipher-locked doors, proximity readers, and motion detectors) to ensure that only personnel who need to access these spaces are granted access.

EXAM TIP Physical security methods use layered defenses. These provide a deterrent control and block or slow down even the most dedicated attackers. A layered defense includes multiple boundaries and can also include multiple physical security protection methods such as guards, motion detectors, and closed circuit television (CCTV) monitors. The most valuable assets, such as systems used in a data center, are placed within the innermost boundary, requiring an attacker to bypass controls for outer boundaries to gain access.

True or false? Alarms are a type of detective control used in physical security.

Answer: *True*. Various types of alarms can be used to detect when a physical security barrier has been breached. In some cases, the alarm can be loud and also act as a deterrent. In other cases, the alarm can alert security personnel without letting the attacker know that the breach has been detected.

Often an organization will combine a strong security design with multiple visible elements of security. For example, security guards and visible CCTV cameras act as deterrent controls discouraging would-be attackers.

Can you answer these questions?

You can find the answers to these questions at the end of the chapter.

1. How many public entry points should a facility include based on the natural access control strategies of CPTED?
2. When designing the location of a server room used as a data center, where should it be placed within a facility?

Objective 10.2: Support the implementation and operation of perimeter security (e.g., physical access control and monitoring, audit trails/access logs)

Perimeter security is the outside boundary of an organization and provides the first layer of physical security for an organization. A primary security method is the use of detection systems that detect unauthorized access and set off alarms. Audit trails and access logs can record all access and help determine when personnel have accessed specific areas. By reviewing failure events in the logs, it's also possible to determine when personnel attempt to access areas where they are not authorized.

Exam need to know...

- Physical access control and monitoring
  For example: What is a volumetric detector? What type of iris should be used for an outdoor camera?
- Audit trails/access logs
  For example: What should be examined when performing a physical access audit? Can proximity badges be used with access logs?

Physical access control and monitoring

Physical security controls are implemented as part of a layered security system to control and monitor access to physical spaces. The controls implemented at the perimeter of an organization control access to the first boundary, which includes the outside area of a building or the entrance area. Detection systems identify when unauthorized personnel enter any area and can be used anywhere within a facility.

True or false? Security guards are a deterrent physical access control.

Answer: *True*. The presence of security guards is often enough to prevent many would-be attackers from attempting an attack.

> **EXAM TIP** Security guards can help ensure that only authorized personnel are allowed access to restricted areas. In some cases, they can be observers and simply ensure that personnel are wearing badges or that personnel are not tailgating without using their credentials. In other cases, they might be active and provide access to an area only after inspecting hand-carried packages and verifying a person's identity.

True or false? Passive infrared (IR) sensors detect when personnel enter a restricted space based on the broken light beam.

Answer: *False*. Passive IR sensors can detect when personnel enter a restricted work space by detecting a change in temperature. When a person enters a room, the warmth of the individual changes the temperature of the monitored area. An active IR sensor uses a light beam and can detect when the light beam is broken.

Physical intrusion detection systems use a variety of different methods to detect the presence of unauthorized personnel. Basic electromechanical systems monitor a circuit and can detect a change in the circuit. For example, a basic window circuit uses a magnetic field between the window frame and the window. Opening the window breaks the magnetic field, triggering an alarm.

> **EXAM TIP** Volumetric sensors use invisible detection fields and can detect changes in the field. For example, a passive IR volumetric detection system can detect changes in temperature from an intrusion. Proximity volumetric detectors create a magnetic field by using capacitance and can detect an intrusion based on changes in this magnetic field. Proximity detectors are often used to protect a specific item, such as a work of art or access to a safe. Basic volumetric motion or movement detectors can detect any type of movement in a room.

True or false? CCTV cameras commonly use a charge coupled device (CCD) to provide clearer images with more details.

Answer: *True*. A CCD is a light-sensitive chip that converts the light into an electrical signal, and some cameras have more than one CCD. In general, a CCD CCTV camera is superior to non-CCD cameras.

CCTV cameras are commonly used as a monitoring method. The presence of CCTV cameras acts as a deterrent control, and when combined with a recording system, they can be used as a detective control. A recording from a CCTV camera often provides incontrovertible evidence of a crime or unauthorized activity.

True or false? A fixed focal length lens on a CCTV camera provides a fixed field of view.

Answer: *True*. Cameras with a fixed focal length lens provide a constant view.

The iris of a camera controls the amount of light allowed into a CCTV camera. Most cameras require an electronic iris that can automatically adjust the size of the iris in response to changing light conditions. Some indoor CCTVs monitoring an area with fixed lighting use a manual iris. The iris is adjusted when the camera is installed, but because the light is constant, it doesn't need to be adjusted again.

> **EXAM TIP** Pan-tilt-zoom (PTZ) cameras allow security personnel to modify the field of view based on activity. Varifocal lenses provide a shallow depth of focus, allowing the viewer to focus on smaller details. IR corrected lenses provide a sharper focus for cameras by focusing visible and IR light on the same focal plane. The iris controls the amount of light allowed into the camera, and if light conditions change, the iris should be automatically adjusted.

> **MORE INFO** Bosch published an excellent paper on CCTV cameras, titled "Selecting the Right CCTV Camera," that covers many of the basics related to CCTV cameras. You can access it by going to their main page (*http://www.boschsecurity.us/en-us/*) and searching "Selecting the Right CCTV Camera." Much of the same information is also available via their Learning Center pages (*http://www.boschsecurity.us/en-us/ProductInformation/cameras/LearningCenter.htm*).

True or false? Perimeter intrusion detection assessment systems include fence disturbance sensors and microwave sensors.

Answer: *True.* A perimeter intrusion detection and assessment system (PIDAS) is designed to detect perimeter intrusions. Fence disturbance sensors detect vibrations or cuts. Microwave transmitters and sensors flood an area with an electronic field and can detect any movement in the field.

> **MORE INFO** The United States Nuclear Regulatory Commission published a comprehensive document titled "Intrusion Detection Systems and Subsystems." While the methods used to protect a nuclear facility might be overkill for most facilities, this document does provide excellent coverage of many basic physical intrusion detection methods. The Exterior Detection Sensors section provides extensive coverage of PIDAS that can be used for any facility. You can access it here: *http://www.nrc.gov/reading-rm/doc-collections/nuregs/staff/sr1959/sr1959.pdf.*

Audit trails/access logs

Audit trails allow you to create a time-stamped record of activity. Technical or logical access control logs use technical means to record events as they occur, such as when a user accesses a file. Physical access control logs identify when personnel access physical areas.

Logs can be handwritten logs entered by a guard. However, these are cumbersome and susceptible to human error. Electronic logs can record access by personnel using technical means and are more reliable. For example, proximity badges used to open doors can be coded to identify individuals who have been issued the badge. Each time a user's proximity badge is used to open a door, the log records whose badge was used to open it and the time when it was opened.

> **EXAM TIP** When auditing physical access, all unsuccessful attempts should be reviewed. Successful attempts are common and show when people enter or exit an area. Unsuccessful attempts show when an unauthorized person tried to access a restricted space. A pattern of a single person trying but failing to access multiple spaces is indicative of a potential problem that needs investigating.

Can you answer these questions?

You can find the answers to these questions at the end of the chapter.

1. What is the purpose of a volumetric sensor?
2. What is the benefit of a CCD in a CCTV camera?
3. When comparing methods to create access logs and record entry and exit to an area, what is the most efficient method?

Objective 10.3: Support the implementation and operation of internal security (e.g., escort requirements/visitor control, keys and locks)

Internal security refers to additional measures taken to protect internal boundaries of an organization. Some primary methods are the use of badges to easily identify authorized personnel and visitors. Electronic access controls such as cipher locks or proximity cards are used to control entry into internal areas.

Exam need to know...

- Escort requirements/visitor control
  For example: What can be used to identify visitors within a restricted area?
- Keys and locks
  For example: What are some types of electronic access controls commonly used to control access?

Escort requirements/visitor control

Restricted areas of an organization must be protected against unauthorized access. An organization will use various methods to restrict access, but occasionally, visitors might need temporary access. A primary method of controlling visitor access is by assigning an escort to stay with the visitor at all times. Escorts should be able to view the visitor at all times.

True or false? Organizations often require employees and visitors to wear badges when walking around employee spaces.

Answer: *True.* Visibly displayed badges worn by personnel provide easy identification of authorized personnel. In most cases, the badges are color-coded or provide some other markings to further identify the access granted to the badge owner. For example, employee badges might be white and visitor badges might be red.

Many organizations require personnel to wear identification badges while inside a restricted or controlled area. Badges will have a picture and other information about the wearer for identification. Unauthorized personnel stand out when they aren't wearing a badge or when wearing a badge with the wrong markings for the area they're in.

> **EXAM TIP** Badges such as common access cards (CACs) and personal identity verification (PIV) cards often serve dual purposes. For example, they can double as smart cards, with an embedded certificate used for authentication on computers and networks. The user will typically be required to enter a personal identification (PIN) or password in addition to using the smart card. In some cases, badges might include proximity electronics used to open a door with a proximity reader, or radio frequency identification (RFID) circuitry to open a door. These can also be paired with a PIN to control access.

MORE INFO Federal Information Processing Standard Publication 201 (FIPS 201), "Personal Identity Verification of Federal Employees and Contractors," identifies the standards for PIVs. SP 800 78-3, "Cryptographic Algorithms for Key Sizes for Personal Identity Verification," identifies the cryptographic standards for PIVs used for authentication. You can access FIPS documents from *http://csrc.nist.gov/publications/PubsFIPS.html*, and you can access SP 800 documents from *http://csrc.nist.gov/publications/PubsSPs.html*.

Keys and locks

Within physical security, keys and locks are used to secure physical resources. While a key implies manual locks, many locks are electronic. This is especially true of door locks that can be opened electronically.

True or false? One of the primary purposes of locks is to delay unauthorized access.

Answer: *True*. Physical security controls typically fall into one of the following categories: deter attacks, delay attacks, detect attacks, and respond to attacks. Locks fall into the delay category, with the goal of delaying an attacker long enough for detection and response mechanisms to stop unauthorized access.

Electronic access control systems are frequently used to control access to an area. They are more effective than a traditional key-based lock. Some are cipher locks, requiring a user to enter a specific cipher code to gain entry. Some use proximity card readers where a user passes a proximity card in front of or in close proximity to the reader. Others use RFID methods to open doors when a user is close or after the user has entered a PIN.

NOTE Cipher locks can be either manual or electronic. A manual cipher lock simply requires personnel to enter the proper code into push buttons and then twist a handle to open the lock. Manual cipher locks do not use electricity. An electronic cipher lock has a numeric keypad, and when the user enters the correct code, the door is opened electronically. In some cases, electronic cipher locks are combined with proximity cards or badges. The user must have a card or badge and must also know the code. The combination of both methods provides stronger access control.

True or false? A fail-safe door will remain locked in the event of a power failure.

Answer: *False*. Fail-safe doors will default to an unlocked state if power fails, which is especially valuable if a fire or other emergency affects the power. That is, the doors are designed to be safe for personnel.

In contrast, some doors with electronic access are designed to fail in a locked position if power is lost. This is also known as fail-secure or fail-soft. For example, a vault would automatically lock or remain locked if power was lost to ensure that the contents remain protected. One way of remembering these differences is by thinking of fail-safe as fail-safety, and fail-secure as fail-security.

Can you answer these questions?

You can find the answers to these questions at the end of the chapter.

1. What physical security method is commonly used to identify authorized personnel within a restricted area?
2. What is commonly included with a PIV card to provide authentication?
3. What state will a fail-safe door be in after a fire results in a power loss?

Objective 10.4: Support the implementation and operation of facilities security (e.g., technology convergence)

A facility's security methods need to provide protection in several different areas. Most people recognize the importance of physical security to protect server rooms and other restricted areas housing information technology (IT) systems. However, other systems, such as heating, ventilation, air conditioning (HVAC) units, and fire suppression units, are also important. If air conditioning fails in a server room, systems can overheat, resulting in hardware failures and long-term outages. Similarly, a fire can be catastrophic if not quickly detected and suppressed.

Exam need to know...

- Communications and server rooms
  For example: What is a danger of unrestricted access to a communication room?
- Restricted and work area security
  For example: What is the primary goal of a clean desk policy?
- Data center security
  For example: What is the difference between data center security and security for a server room?
- Utilities and heating, ventilation and air conditioning (HVAC) considerations
  For example: What are hot and cold aisles?
- Water issues (e.g., leakage, flooding)
  For example: If an organization operates in a flood plain, what should it include in its facility?
- Fire prevention, detection and suppression
  For example: What should never be used on an electrical fire? What type of cables should be used in plenums?

Communications and server rooms

An organization will commonly dedicate a specific area for communications equipment and servers. This allows the organization to provide additional security for this equipment and limit the access. In many cases, communications equipment used to access phone lines or provide a direct interface to the Internet are colocated in a server room or located separately in a wiring closet.

True or false? The best location for a wiring closet hosting communications equipment is on an outside wall.

Answer: *False*. While locating a wiring closet on an outside wall might be convenient for connections, it doesn't provide the best security. It is best to locate a wiring closet within an internal boundary of the organization.

In addition to hosting servers and other communications equipment, a server room will usually include networking components such as routers and switches. In small organizations, these networking components might be colocated in a wiring closet with the communications equipment. However, no matter where they are located, these networking components must be protected with physical security measures to restrict access.

> **EXAM TIP** If an attacker is granted unauthorized access to a router or switch, it's relatively easy to connect a wireless access point to the device. Unauthorized access points used in this way are called rogue access points. The access point can capture all the data going through the network and transmit the contents wirelessly. If the attacker uses a directional antenna instead of a standard omnidirectional antenna to capture the traffic, it's possible to capture the intercepted traffic from more than a mile away. This allows the attacker to passively collect the data without any personal risk after the access point is installed.

Restricted and work area security

Access to any restricted or work areas is controlled to prevent unauthorized access. The basic method of controlling access is by using badges and electronic access control systems.

True or false? Employees should be granted access to spaces based on only the rights and permissions they have been granted.

Answer: *True*. Physical access to specific areas should be based on privileges assigned to personnel. If personnel do not have privileges for IT resources within a restricted area, they should not be granted physical access. This follows the standard security principle of least privilege.

Social engineering techniques are also a concern when considering work area security. For example, shoulder surfing is the practice of looking over a person's shoulder or observing what is displayed on their monitor.

> **EXAM TIP** Shoulder surfing techniques allow unauthorized personnel access to information simply by looking at the display screen. Password masking displays a special character, such as *, instead of the actual password to mitigate a shoulder surfer's ability to collect passwords. Additionally, display screen filters can restrict the viewing angle of a monitor to reduce the ability of unauthorized personnel to view the display.

Many organizations implement clean desk or clear desk policies. These reduce the amount of useful material that is available if an unauthorized individual gains access to an employee's desk while the employee is away. It also helps to present a positive image of a company when customers visit.

> **MORE INFO** CSO Online published a good article about clean desk policies that you can view here: *http://www.csoonline.com/article/219055/*. It starts with a picture that enables you to identify security issues with a desk and follows it with clear explanations of common issues.

Data center security

A data center houses a large number of servers and related computer systems. The purposes vary, but some common uses of data centers are as network operations centers (NOCs) providing centralized management of a large network, specialized facilities housing websites and supporting services for outside companies, and large server centers used to provide cloud-based services for internal or external clients.

True or false? Data centers include highly redundant systems, so formal disaster recovery plans are not necessary.

Answer: *False*. A data center includes critical business functions and systems and needs a disaster recovery plan. The business continuity and disaster recovery plans are often more stringent and managed more aggressively for a data center than for other areas of a business. It's also true that a data center will have more fault-tolerant systems to eliminate single points of failure and increase overall availability of systems.

> **EXAM TIP** The value of the data and services within a data center is typically at the top of an organization's asset list. The potential losses if the data is lost or the services fail is significant. Because of this, the physical and environmental security required for a data center is typically a combination of the strongest physical and environmental controls used for any facility.

> **MORE INFO** Sean Heare wrote a Data Center Physical Security Checklist, which is available in the SANS Institute InfoSec Reading Room. It provides a comprehensive list of physical security elements used to protect data centers and also provides a good review of the entire physical security domain. You can access it here: *http://www.sans .org/reading_room/whitepapers/awareness/data-center-physical-security-checklist_416*.

Utilities and heating, ventilation, and air conditioning (HVAC) considerations

HVAC systems are an integral part of any organization that houses computer systems. They provide a consistent temperature and control the humidity, which both contribute directly to overall availability. Electronic components can overheat and fail when the temperature gets too hot, and other problems occur when the humidity isn't controlled.

True or false? Hot and cold aisles are commonly used to provide better temperature regulation in a data center.

Answer: *True*. Hot and cold aisles are used in server rooms and data centers to control air flow.

Typically air flows from the bottom or the front of a bay through the back or top of the bay. A server room has multiple rows of bays, and if all the bays are facing the front of the server room, hot air exhausted from one bay is brought into the front of the bay behind it. The further you go to the back of the room, the hotter the air becomes.

In a hot and cold aisle design, the bays are organized so that the back of each row of bays faces the back of the next row. Similarly, the front of each row of bays faces the front of the next row. Cold air is forced through the floor of the aisles with the front of the bays (called the cold aisles) and sucked in through the front of the bays. Hot air comes out of the rear of the bays into the hot aisles and is directed toward HVAC input ducts.

> **MORE INFO** For a short review, with some graphics, of how hot and cold aisles can be effectively designed, check out this page on 42u's website: *http://www.42u.com/cooling/hot-aisle-cold-aisle.htm*. The site also includes a link to a more detailed recorded webinar.

True or false? Humidity within a server room should be kept as low as possible.

Answer: *False*. Ideally, humidity should be around 50 percent. When the humidity is too low, it increases the risks related to electrostatic discharge (ESD). When the humidity is too high, it increases the risks related to condensation and moisture.

Water issues (e.g., leakage, flooding)

Water sensors are used to detect water from either leaks or flooding. Water sensors are especially important for an organization in an area prone to flooding but are also used elsewhere. For example, water sensors can detect flooding from a burst pipe within the building.

It's common to connect water sensors to a pumping system. When the water is detected, pumps are automatically activated to pump the water out to prevent damage.

Fire prevention, detection, and suppression

A fire can quickly consume a facility and all of an organization's assets, making it a very serious risk. However, there are many basic steps an organization can take to mitigate the risk. Basic prevention steps help prevent a fire from starting at all. Detection methods are designed to discover a fire as quickly as possible, and suppression methods attempt to extinguish it as soon as it is detected.

True or false? Only Class A fire extinguishers should be used on electrical fires.

Answer: *False*. Class A fire extinguishers use water and can be fatal if used on electrical fires. Only fire extinguishers designated for electrical fires should be used to extinguish an electrical fire.

The primary ingredients required for a fire are oxygen, heat, and fuel. Chemical chain reactions can start or sustain a fire if the other ingredients are present. Fire suppression techniques attempt to block one of these elements to extinguish a fire.

Many data centers and IT facilities are equipped with gaseous-based fire suppression agents. Ideally, these gases will starve the fire by removing one of the elements of the fire without harming personnel or equipment.

Gases such as MH227 and FM-200 are often used in facilities because they can extinguish a fire but represent minimal risk to personnel and equipment. Carbon dioxide (CO_2) has been used in the past, but because it reduces the amount of available oxygen, it is harmful to personnel. Halon quickly stops the chemical reaction but is dangerous to the environment and personnel and is outlawed in many countries.

Table 10-1 shows the classifications for fires in different countries/regions.

TABLE 10-1 Fire Classifications

TYPE OF FIRE	NORTH AMERICA	EUROPE	AUSTRALIA/ ASIA
Ordinary combustibles	Class A	Class A	Class A
Flammable liquids	Class B	Class B	Class B
Flammable gases	Class B	Class C	Class C
Electrical equipment	Class C	Class F/D	Class E
Combustible metals	Class D	Class D	Class D
Cooking oil or fat	Class K	Class F	Class F

EXAM TIP It is important to remember that water conducts electricity and should never be used on an electrical fire. Gas-based suppression agents such as FM-200 are often used in IT facilities to extinguish electrical fires.

True or false? Cables rated as plenum-safe are used to prevent the spread of fires and toxic fumes.

Answer: *True*. Cables that go through plenums (spaces used for heating and air conditioning) must be rated as plenum-safe. The jacket on these cables is fire-retardant to prevent the spread of a fire through a plenum. Additionally, the jacket does not emit toxic fumes if it burns or melts.

There are a variety of different types of detectors used to detect a fire. Smoke detectors are the most common and are based on the premise "where there's smoke, there's fire." Most smoke detectors use optical sensors to detect the smoke. Photoelectric detectors work similarly but instead send a beam of light between two sensors and can detect when smoke particles break the beam of light, similar to how an active IR detector can detect an intrusion when a light beam is broken. Heat-based detectors will cause an alarm when the temperature exceeds a certain temperature or when the temperature increases quickly.

Can you answer these questions?

You can find the answers to these questions at the end of the chapter.

1. What is the risk of a rogue access point in a communications room?
2. What is a physical security method that can be used to prevent shoulder surfing?
3. What is the difference in security requirements for a server room and a data center?
4. What is the risk of allowing the humidity level to drop too low in a server room?
5. You are implementing a gas-based fire suppression system in a server room occupied 24 hours a day. Given a choice between carbon dioxide and FM-200, what is the best choice?

Objective 10.5: Support the protection and securing of equipment

The major threats to equipment are theft, damage, and unauthorized access. Unauthorized access to equipment is largely blocked by implementing various physical security controls covered previously in this chapter. However, there are some additional steps that can be taken to reduce theft and damage.

Exam need to know...

- Prevent the theft of equipment
  For example: What can users do to prevent the theft of laptops? What can prevent damage to computers from commercial power anomalies?

Asset inventory systems provide an effective means for an organization to track valuable assets. Many organizations use barcode or RFID database systems to identify and track their valuable assets. They then perform periodic inventories to verify that the assets have not been stolen.

True or false? Training users about physical security can reduce the incidence of laptop thefts.

Answer: *True*. Laptops are frequently stolen while users are traveling or attending training sessions away from the company. When users understand the common ways laptops are stolen and are taught simple methods to prevent the thefts, it significantly reduces the risk.

Cable locks that connect to a laptop and a piece of furniture are effective at reducing thefts. These work similar to a cable lock used to lock a bicycle to a bicycle ramp. Additionally, hand-carrying a laptop onto an airplane instead of checking it as luggage is strongly recommended. Checked luggage is often thrown around, which can cause damage. Additionally, a checked laptop is unattended at many times during a trip and can be easily stolen.

True or false? Surge protectors ensure that systems have consistent clean power.

Answer: *False*. Surge protectors protect only against spikes and surges but do not regulate power. Power line conditioners or regulators can ensure that systems have clean power.

Providing consistent clean power to equipment is an important concern when addressing availability. At a minimum, all systems should be protected with basic surge protectors to prevent spikes and surges from reaching equipment. Power line conditioners and regulators can be used when the commercial power includes interference or is not a consistent voltage. It will strip off all interference and regulate the output at a steady voltage.

EXAM TIP An uninterruptible power supply (UPS) provides short-term power for a system when commercial power is lost and usually includes surge, spike, and sag protections. When a line conditioner is used with an UPS system, the UPS is placed between the line conditioner and the equipment. The line conditioner provides clean power to the UPS system. The UPS then protects the computer system from power fluctuations, including short-term power failures.

Can you answer these questions?

You can find the answers to these questions at the end of the chapter.

1. What can users do to prevent the theft of a laptop?
2. What are different methods used to prevent damage to computer components from power anomalies?

Objective 10.6: Understand personnel privacy and safety (e.g., duress, travel, monitoring)

The last objective in this domain addresses personnel privacy and safety. There is some overlap in personnel privacy issues with other domains, reflecting the importance of protecting personnel privacy and personally identifiable information (PII). Safety of personnel is also important, and by training personnel on various safety issues related to security, it helps them avoid problems when away from the work environment.

Exam need to know...

- Personnel privacy and safety
  For example: What can a traveler do before a trip to reduce risks? How can a laptop be protected against unauthorized data access?

Personnel privacy and safety

Personnel privacy issues primarily focus on the protection of PII. It's as important to protect this information for internal employees as it is for external customers, and the requirement to do so is commonly governed by laws.

> **MORE INFO** NIST SP 800-122 is titled "Guide to Protecting the Confidentiality of Personally Identifiable Information (PII)," and it provides detailed definitions of PII and methods to protect it. It can be downloaded from the following NIST document page: *http://csrc.nist.gov/publications/PubsSPs.html.*

True or false? Training personnel on safety and security issues after a trip is often useful in helping them avoid problems.

Answer: *False*. Training personnel before a trip on specific issues related to their destination is useful in helping them avoid problems. Training them after the trip is too late. Common actions people can take to protect themselves while on a trip are to remain vigilant and to be aware of their surroundings. Certain locations are known for specific threats, and being aware of the current threats makes a person less likely to become a victim.

Basic protections include not inviting a stranger into your room, not leaving drinks unattended, being wary of new acquaintances asking many questions, and being aware that your conversations might be monitored. There are a wide variety of scams and cons that criminals use, but when travelers are aware of the criminals' techniques, travelers are less likely to be tricked.

> **MORE INFO** The U.S. Federal Bureau of Investigation (FBI) has published a brochure titled "Safety and Security for US Students Traveling Abroad." While the title indicates that it's only for students, the tips apply to anyone traveling overseas for work or pleasure. You can access it here: *http://www.fbi.gov/about-us/investigate/counterintelligence/student-travel-brochure-pdf.*

When using mobile computers in a public place, a basic premise is to be suspicious of all public connections. It's very easy for an attacker to set up a wireless network that looks legitimate but is malicious. For example, an *evil twin* is an access point with the same name as a legitimate access point.

Firewalls and antivirus software should be enabled, up-to-date, and using the most secure settings for a public location. Many host-based firewalls include a setting that blocks all unsolicited connections from a public network.

If you can do without electronic devices such as laptops, tablets, and smartphones, it's best to leave them behind. Additionally, personnel should remove all sensitive data from their devices to protect them from accidental theft. There are many stories of how data has been taken from user systems while users traveled overseas.

Can you answer these questions?

You can find the answers to these questions at the end of the chapter.

1. What should a traveler do if someone knocks on their hotel room claiming to be room service, even though room service has not been requested?
2. What is the most effective method of ensuring that data is not stolen from a user's system when traveling overseas?

Answers

This section contains the answers to the "Can you answer these questions?" sections in this chapter.

Objective 10.1: Understand site and facility design considerations

1. A building should have one public entry point. An organization will often have additional entry points for employees, but these will have different entry controls than the public entry.
2. A server room should be placed within other boundaries, so it is typically closer to the center of a building. This requires an attacker to breach all other boundaries before reaching the server room.

Objective 10.2: Support the implementation and operation of perimeter security (e.g., physical access control and monitoring, audit trails/access logs)

1. Volumetric sensors are used to detect intrusions. They use invisible detection fields to monitor an area and can detect changes in the area. For example, when a person enters an IR monitored field, an IR sensor can detect changes in temperature and set off an alarm.
2. The CCD is a light-sensitive chip within a CCTV. It converts the light into an electrical signal, and cameras with one or more CCDs provide clearer images, with more details than cameras without a CCD.
3. Technical methods used to create and record access are far superior to any manual method. This includes using security guards to record access because security guards are susceptible to human error.

Objective 10.3: Support the implementation and operation of internal security (e.g., escort requirements/visitor control, keys and locks)

1. Badges are commonly used to easily identify authorized personnel. In many cases, the badges are color coded or have other markings to identify different types of access.
2. CACs and PIVs often include certificates, allowing them to function as smart cards for authentication in addition to being used as identification badges.
3. A fail-safe door fails in an open state after a power failure. This provides safety for personnel and ensures that they have the ability to exit an area in the event of a fire or other emergency.

Objective 10.4: Support the implementation and operation of facilities security (e.g., technology convergence)

1. A rogue access point is an unauthorized wireless access point. If it is connected to a router or a switch in a communications room, it can transmit network traffic to an attacker located outside the facility's physical boundaries.
2. Display screen filters can be placed over a monitor to limit the viewing angle and reduce the ability of an attacker to read information. Password masking can also be used to reduce the risk of someone reading a password as it's typed, but this is a technical method, not a physical method.
3. In general, the security requirements for a data center and a server room are similar. The majority of the security requirements will be the same. However, a data center might house higher-value assets than a server room and might include more stringent security requirements.
4. If the humidity level is too low, it increases the potential for damage from electrostatic discharge. The humidity level should be around 50 percent.
5. FM-200 is the best choice because it presents the least risk to personnel and equipment. Carbon dioxide will extinguish the fire with minimal damage to the equipment, but it is harmful to personnel.

Objective 10.5: Support the protection and securing of equipment

1. Users can prevent the theft of laptops by not leaving them unattended. Cable locks are useful to physically secure the laptop when it is not attended.
2. Power line conditioners or regulators can remove interference and regulate power, although these aren't needed very often. An UPS includes a battery and provides continuous voltage, even when power is lost. A surge protector protects against spikes and surges, and many UPS systems commonly include surge protection capabilities.

Objective 10.6: Understand personnel privacy and safety (e.g., duress, travel, monitoring)

1. It's best not to invite strangers into your room no matter who they say they are. If they claim to be providing room service but room service wasn't ordered, you can call the front desk to verify their identity or request security assistance.
2. The only way to guarantee data is not stolen from a system is to not bring the system on the trip. If a user brings a system, it should be sanitized by removing all valuable data, including all remnants of the data, from the system.

Index

Symbols

802.11a wireless protocol, 32
802.11b wireless protocol, 32
802.11g wireless protocol, 32
802.11n wireless protocol, 33
*-Integrity Axiom (star Integrity Axiom), 146
*-property (star property) rule, 5, 146

A

academic software, 221
acceptable use policy (AUP), 2, 76
access aggregation, 15–16
access control
    assessing effectiveness, 17–19
    basics, 1–12
    effective practices and accountability, 231
    natural, 248
    non-discretionary or discretionary, 5
access control attacks, 12–16
Access Control domain, 1
access control lists (ACLs), 10, 34
access control matrixes (ACM), 10
access control policy, 170
access control strategy, 2
access reviews, 18–19
accountability, effective access control practices and, 231
account lockout policies, 20, 21
accounts. *See* user accounts
ACID model, 97
ACK packet, 29
acrylic glass, 249
Active Directory Certificate Services (Microsoft AD CS), 136
Active Directory (Microsoft), 7
active IR sensor, 250
active response by IDS, 181
ActiveX controls, 98
Adaptive Data Storage (ADS), 204
Address Resolution Protocol (ARP), 26
administrative accounts, 17, 18
    revocation of account access, 22
administrative controls, 3

administrative-level permissions, 169
Advanced Encryption Standard (AES), 7, 33, 96, 106, 111, 112
Advanced Persistent Threats (APTs), 16, 219
advisory security policies, 60
aggregation attacks, 161
AIC triad, 56–58
airplane mode for mobile device, 35
alarms, 249
American Standard Code for Information Interchange (ASCII), 27
annual loss expectancy (ALE), 72
annual rate of occurrence (ARO), 72
anomaly-based detection, 180
anonymous relays, 180
antivirus software, 181–182
    on mobile computers, 262
Apple devices, location data stored on, 238
Application layer (OSI), 27
application level firewalls, 34
applications
    cryptography for security, 132–134
    patch management, 182–183
arc 4 encryption algorithm, 112
assessment, in disaster recovery process, 209
assets
    inventory systems, 260
    management, 174
    valuation, 14–15
        tangible and intangible, 73
asymmetric cryptography, 113–115
Asynchronous Transfer Mode (ATM), 30
atomicity, in ACID model, 97
attackers, error message information and, 95
attacks, 178–180
    cryptanalytic, 124–129
    preventive measures for, 180–182
audio file, hiding data in, 139
audio over IP networks, 37
audit committee, 50
audits, 18, 240–241
audit trails, 11, 231, 252
    for user account changes, 17

authentication, 6–8
    and confidentiality, 56
    digital signatures and, 121
    FFIEC on, 62
    SSL and TLS for, 39
authentication, authorization, and accounting (AAA) services, 8
Authentication Header (AH), 39, 132
authorization, 9–24
automated inventory systems, 174
automated processes to deploy updates, 183
availability, 56
    data in cloud computing, 244
    fault tolerance and, 191
    protecting, 48
    of resources, 58
awareness of employees on security, 78–79

B

backdoors, 99, 101
background check of job candidates, 74–75
backups
    metrics on, 81
    storage strategy, 202–204
    tapes as security resource, 82
badges, 253
    proximity, 252
barcode database systems, for assets, 260
baseline, 60–61
    for computer configuration, 187
bastion host, 34
beacons, 159
Bell-LaPadula model, 5, 6, 146
Biba model, 5, 6, 146
biometrics, 6, 7
birthday attack, 126
bit-level copies, 237
bit-level image, 233
BitLocker Drive Encryption (Microsoft), 106, 155
bits, 28
black box test, 184
block ciphers, 112
Blowfish, 112
bluejacking, 40
Bluetooth, 40
Bosch, on CCTV cameras, 251
bot herders, 42, 43, 156, 236
botnet servers, 236
bounds and boundary checking, 101

Brewer-Nash model, 6, 147
browsewrap contract, 222
brute force attacks, 126
budget, for security, 80–88
buffer overflow, 13, 99
business continuity, 195
Business Continuity & Disaster Recovery Planning domain, 197
Business Continuity Management Institute (BCM Institute), 204
business continuity plan (BCP)
    creation of, 196
    exercises and reviews for testing, 211–214
    focus of, 196
    maintaining, 213
    project scope development and documentation, 196–197
    requirements, 195–198
business impact analysis (BIA), 195, 198–202
bus network topology, 28
bytes, 139

C

cable locks, 260
cables, 31, 32–33, 157
    disconnecting from network to protect evidence, 230
    plenum-safe, 259
call tree for personnel notification of disaster recovery response, 207
cameras, CCTV, 251
capability maturity model (CMM), 89, 92
capability table, 10
carbon dioxide (CO_2), 259
carbon footprint, 154
Carnegie Mellon
    CERT note, 42
    Software Engineering Institute, 92
carrier, for steganography, 140
CCTV cameras, 251
certificate authority (CA), 135–136
    private, 136
certificate revocation list (CRL), 138
certificates, 133, 137–139
    public keys in, 113
    validation of, 138–139
certificate trust chain, 136
chain of custody log, 230–231
change advisory board (CAB), 186

change management, 93–94, 185–186
Chargen protocol, 42
checkpoint process, 97
checksum, 116
chief financial officer (CFO), 53
children, parental consent for website information collection, 225
Children's Online Privacy Protection Act (COPPA), 225
Chinese Wall model, 6, 147
chosen ciphertext attack, 128
chosen plaintext attack, 125
CIA triad, 56–58
Cipher Block Chaining (CBC), 113
ciphertext data, 110, 111
ciphertext-only attack, 127
Cisco
  DocWiki site, 28
  guide to ATM technology, 30
  NetFlow, 236
CISSP certification. *See also* exam need to know
  code of ethics and, 226
civil laws, 239
Clark-Wilson model, 6, 147
Class 1 certificate, 137
Class A fire extinguishers, 259
classification of information, senior management responsibility for, 64
clean desk policies, 257
clean power, 261
click fraud, 219
clickwrap contract, 222
client-based software, threat against, 159–160
client-side scripts, 98
cloud computing, 162, 242
  data protection and backup, 242
  solutions for business continuity, 206
code. *See* source code
Code of Ethics, 225–227
cold sites, 204
collection of evidence, 230–231
collisions, 127
committed transactions, 97
Committee of Sponsoring Organizations for the Treadway Commission (COSO), 54
common access cards (CACs), 253
Common Criteria, 149–153
communication channels, 36–40
communications
  in disaster recovery process, 208–209
  physical security for equipment, 255–256
compartmentalized environment, for label assignment, 5
Compartmented security mode, 148
compensating controls, 4
computer crime, 217, 218–219
concurrency controls, 97
confidentiality of data, 56–57, 65, 110
  with cloud computing, 243
  from email cryptography, 133
  protecting, 48, 105
configuration management, 61, 100, 185, 186–187
consistency, in ACID model, 97
constrained user interface, 9
consultants, 77
containment of security incident, 176, 230
content-dependent access controls, 161
context-dependent access controls, 161
continuity of operations plan (COOP) sites, 204
contractors, 77
contracts
  ensuring security in, 242
  for software, 222
  with vendors, 242
Control Objectives for Information and related Technology (COBIT), 53–54
copyright, 219, 220
  for software, 221
  laws, 162
corrective controls, 4
cost
  of backups, 82
  of countermeasure, 80
cost-benefit analysis, 14, 80
counterattacks, 229
countermeasures
  cost of, 80
  in risk management, 72–73
  principles, 163
covert channels, 156
C programming language, 95
creative works, copyrights to protect, 220
credit card companies, PCI DSS standard for, 61, 149, 151, 240
crime prevention through environmental design (CPTED) strategies, 247
criminal laws, 239

critical business functions, identifying and prioritizing, 198–200
crossover error rate (CER), 7
cross-site script (XSS), 44, 98
    preventing attacks, 101
cryptanalytic attacks, 124–129
cryptographic keys, 110, 111
    creation/distribution, 119
    key escrow, 120
    storage/destruction, 120
cryptographic life cycle, 107–109
cryptography
    algorithm/protocol
        governance, 108–109
    application and use, 105–107
    asymmetric, 113–115
    basics, 109–118
    for application security, 132–134
    for network security, 129–132
    key transmission, 112
    symmetric, 111–113
Cryptography domain, 105
CSO Online, 257
customers, communication about disaster, 209

D

damage assessment team, 208
data at rest, encryption, 106
databases
    access control for, 147
    deadlock, 161
    security, 160–161
    transaction logs, 97
    views as constrained user interface, 10
Data Center Physical Security Checklist, 257
data center, security, 257
data communications, 39–40
data custodians, 51
data emanation, risk of, 157
Data Encryption Standard (DES), 7, 112
data flow, trans-border, 223–224
datagrams, 28
data integrity, SSL and TLS for, 39
data in transit, 106–107
data in use, 153
data leakage, 39, 160
Data Link layer (OSI), 28
data-loss prevention devices, 160
data-loss prevention software, 40

data marts, 97
data mining, 97
data owners, 51, 64
data warehouse, 160
date stamps, 95–96
deadlock, 97, 161
decryption process, 110
Dedicated security mode, 148
defense in depth, 163
defense-in-depth strategy, 35
degaussing, 171, 172
deleting file, vs. sanitization, 171
demilitarized zone (DMZ), 34
denial of service (DoS) attack, 13, 41, 178
    application failure from, 96
    attacker identification, 43
Department of Defense (DoD) model, 26
design
    in development life cycle, 90
    of sites and facilities, for security, 247–249
detection systems, 250–252
detective controls, 4
deterrent controls, 4
dial-up connections, 38
Diameter, 8
differential backups, 202
Diffie-Hellman algorithm, 115
digital signatures, 121–123
    email validation with, 133
    videos related to, 123
digital voice communications, 36
directive controls, 4
disabled accounts, 22
disaster recovery plan, for data center, 257
disaster recovery strategies, 202–206
    development and
        documentation, 196–197
    recovery process, 206–211
        assessment, 209
        communications, 208–209
        employee training, 211
        personnel, 208
        response, 207
        restoration, 209–210
    recovery site strategies, 204–206
disasters, impact of, 197
discretionary access controls, 5
Discretionary Security property rule, 146
Discretix Technologies, Ltd., 129
disks, fault tolerance for, 188–190

display screen
  filters, 256
  harvesting data from, 39
disposition/disposal, in development life cycle, 91
disruption in service, 195, 199
distributed computing environments (DCEs), 8–9
distributed denial of service (DDoS) attack, 13, 41, 42, 178
  attacker identification, 43
distributed systems, 162
DNS-related attacks, 14
DNS spoofing, 43
documentation
  in change management, 185
  of exercise testing BCP, 213
  for identifying vulnerabilities, 68
  and regulation compliance, 66–67
  for security incidents, 231–232
  of security policies, 62–63
Domain Name System (DNS), 27
domains, 10
drive-by downloads, 98
drug testing, 75
due care, 55
due diligence, 55
  in hiring practices, 75
dumpster diving, 13
duplexing, RAID-1 with, 188
durability, in ACID model, 97
Dynamic Host Configuration Protocol (DHCP) logs, 236
dynamic ports, 29

E

Easter eggs, 101
eavesdropping, video conferencing and, 37
education of employees on security, 78–79
electrical fires, extinguishers for, 259
electromagnetic interference (EMI), 32
Electronic Code Book (ECB), 113
electronic data, movement across government boundaries, 223–224
electronic logs, 252
electronic vaulting, 203
El Gamal algorithm, 115, 128
elliptic curve cryptography (ECC), 115
email
  encrypting and sending, 134

privacy limitations, 53
  secure, 133
  targeting executives with phishing, 159
email servers, 180
emergency management team, 208
employees. *See* personnel
EmployeeScreenIQ, 75
employment agreements and policies, 75–76
Encapsulating Security Payload (ESP), 39, 132
encapsulation, 26
EnCase, 237
enclosures, shielded, 157
encryption, 36, 39, 40, 56, 105
  algorithm, 111
  basics, 109–118
  data at rest, 106
  with hashing algorithm, 6
  link vs. end-to-end, 129
  for VPNs, 38
end-point security, 35
end-to-end encryption, link encryption vs., 129
end user license agreement (EULA), 222
environment
  and security controls, 94–100
  regulatory, 239–240
ephemeral ports, 29
E-Privacy Directive, 52, 224
equipment protection and securing, 260–261
error handling, 101
error messages, 95
escort requirements for visitors, 253–254
escrow of cryptographic keys, 120
ethics code, 225–227
European Union, 52
European Union Directive 2002/58/EC, 224
Evaluation Assurance Level (EAL), 150
evidence
  collection, 176
  collection and handling, 230–231
  corruption risk from combining data from multiple sources, 234
  protecting, 230
evil twin, 33, 262
exam need to know
  on access control, 2, 17
  on access control attacks, 12–13
  on business impact analysis, 198
  on certificates, 137

exam need to know (*continued*)

on code of professional ethics, 225
on communication channels, 36
on contracts, 242
on cryptanalytic attacks, 125
on cryptographic life cycle, 108
on cryptography basics, 109–110
on digital signatures, 121
on disaster recovery process, 206
on facilities security, 255
on forensic procedures, 233
on internal security implementation, 253
on investigations, 228
on key management, 118
on legal issues in international information security, 218
on network attacks, 41
on network components, 31
on non-repudiation, 124
on perimeter security, 250
on personnel security, 74
on provisioning life cycle, 19
on regulations compliance, 239
on risk management, 67–68
on secure network architecture, 26
on security architecture vulnerabilities, 155
on security function management, 80
on security governance, 49
on security incidents, 174–175
on security models, 146
on security operations, 168
on security policy development and implementation, 59
on software and system vulnerabilities and threats, 158
on software development, 90
on software environment, 94
ex-employees, policy on accounts of, 18
exercises for testing BCP, 211, 212–213
exit interview user accounts disabled during, 76
expiration of contractor accounts, 77
exporting goods and services, 222
exposure factor from risk, 70
Extensible Access Control Markup Language (XACML), 9
Extensible Markup Language (XML), 158
standards based on, 9
external companies, data protection by, 244

extract, transform, load (ETL) techniques, 160

F

facilities. *See also* physical security
factors of authentication, 6
failover clusters, 187, 190
    Microsoft server support for, 191
fail-safe door, 254
fail-secure, 254
fail-soft, 254
false positives, from IDSs, 180
Faraday cage, 156
fault tolerance, 187
    and backups, 203
    for disks, 188–190
    for servers, 190–191
Federal Deposit Insurance Corporation (FDIC), 170
Federal Desktop Core Configuration (FDCC), 61
Federal Financial Institutions Examination Council (FFIEC), 62
Federal Information Processing Standard Publication 201 on standards for PIVs, 254
federated identity, 9
    management systems, 8
fence disturbance sensors, 252
fiber optic cable, 32, 157
fiduciary relationship, 55
File Checksum Integrity Verifier (FCIV), 117
File Transfer Protocol (FTP), encrypting traffic, 106
fingerprinting, 7, 102
fire, classifications for, 259
fire prevention, detection, and suppression, 258–260
firewalls, 33–34
    deny-all rule, 148
    filtering traffic at, 43
    on mobile computers, 262
fixed focal length lens on CCTV camera, 251
Flash animations, 98
FM-200, 259
focal length lens on CCTV camera, fixed, 251
foreign key in database table, 97
forensic analysis procedures, 231
forensic collection procedures, 231

forensic procedures, 233–238
    hardware/embedded device
        analysis, 237–238
    media analysis, 233–235
    network analysis, 235–236
    software analysis, 236–237
forensic toolkit, 237
fraggle attack, 42
Frame Relay WANs, 30
fraud
    framework to reduce, 54
    preventing, 167–173
    reducing risk with job rotation, 170
freeware, 221
frequency analysis attack, 128
full backup, 202
full-scale exercises, 212
functional exercises, 212
functional security policies, 60

G

garbage collection, 153
Generic Security Service Application Programming Interface (GSS-API), 95
glass for windows, 248–249
global positioning system (GPS), 35, 238
goals of organization
    security function and, 48
    security policies for, 59
goodwill, 73
    risk of loss, 176
governance
    for security, 49–56
    third-party, 65–67
Gramm-Leach-Bliley Act (GLBA), 244
grandfather-father-son backup strategy, 203
graphic file, hiding data in, 139
graphics processing units (GPUs), 108
gray box testing, 184
grouping subjects and objects, 5
Group Policy (Microsoft), 187
groups, monitoring membership, 22
guidelines, 62
    standards vs., 60

H

halon, 259
hardware components, 31
hardware/embedded device
    analysis, 237–238

hash, 115
hashed message authentication code (HMAC), 117, 123
hashing, 57, 110, 111, 116–118, 235
    algorithms for steganography detection, 140
    definition, 96
    SHA-1 and SHA-256 for, 40
Health Insurance Portability and Accountability Act (HIPAA), 66, 239–240
Heare, Sean, 257
heat-based detectors, 260
heating, ventilation and air conditioning (HVAC), 257
hierarchical environment, for label assignment, 5
high performance computers (HPCs), U.S. regulation of exports, 223
hiring, background check before, 74–75
honeypots, 43, 179
host-based firewalls, 34
hot and cold aisles, 58, 258
hot sites, 204, 205
human threats, 68
humidity, in server room, 258
hybrid cryptography, 115
hybrid risk assessment, 69
hyperlink spoofing attack, 43
Hypertext Transfer Protocol (HTTP), 27
Hyper-V, 154

I

identification, 6–8
identity theft, 224
implementation attacks, 128
implementation, in development life cycle, 91
implicit deny rule, 9, 33
import/export, 222
incident response teams, 228
incremental backups, 202
Incremental model, 91
inference attack, 161
information classifications, senior management responsibility for, 64
information hiding, 139–140
information life cycle, 63–65
information security
    international legal issues, 217
    strategies, 83

Information Security Governance & Risk
 Management domain, 47
information systems
 evaluation models for security, 149–153
 security capabilities of, 153–155
Information Systems Audit and Control
 Association, 54
Information Technology Infrastructure
 Library (ITIL), 152, 186
informative security policies, 60
infrared (IR) corrected lenses, 251
infrared (IR) sensors, 250
Infrastructure as a Service (IaaS), 243
initiation, in development life cycle, 90
input validation, 99, 101
instant messaging (IM), 37–38
intangible assets, valuation, 73
integration
 in development life cycle, 90
 testing, 101
integrity of data, 56, 57, 95
 with cloud computing, 243
 digital signatures and, 121
 hashing to verify, 116
 protecting, 48
intellectual property, 217, 219–222
internal security, implementation and
 operation, 253–255
International Data Encryption Algorithm
 (IDEA), 112
international legal issues in information
 security, 217
International Telecommunications Union
 (ITU), 137
Internet Architecture Board (IAB), RFC 1087,
 on ethics and Internet, 227
Internet Assigned Numbers Authority
 (IANA), 29, 34
Internet Control Message Protocol
 (ICMP), 34
Internet Protocol (IP) networking, 29
Internet Protocol security (IPsec), 38, 39,
 107, 132
Internet Relay Chat (IRC), 156
Internet Security Association and Key
 Management Protocol (ISAKMP), 39
Internetworking Technology Handbook, 28
interrogation, 231
intrusion detection systems (IDSs), 11–12,
 43, 181
 report, 81
intrusion prevention systems (IPSs), 11–12,
 43, 180, 181
inventory systems, automated, 174
investigations of security incidents, 228–232
 protecting sensitive information in
 reports, 232
 roles and responsibilities, 228–229
Invocation property rule, 146
IP addresses
 identifying attack source, 236
 spoofed, 229
IPv4, 29
IPv6, 29
 vulnerabilities, 156
iris scanners, 7
Ironport Systems, 40
ISACA, 54
$(ISC)^2$, 225
 Code of Professional Ethics, 226–246
ISO/IEC 15408, 149
ISO/IEC 27002, 152
isolation
 in ACID model, 97
 of attacked system, 230
IT control framework, 53
IT Governance Institute, 67

J

Java, 95, 98
JavaScript, 98
job candidates, background check of, 74–75
job rotation policies, 2, 170

K

Kerberos, 7, 119
 time stamp use, 96
key derivation function (KDF), 119
Key Distribution Center (KDC), 119
key management, 118–121
 recovery, 120
keys and locks, 254
Knowbe4.com, 83
known plaintext attack, 127

L

labels
 for data, 171
 in mandatory access controls, 5
landscaping, 248

laptops, training users to prevent
   theft, 260–261
Layer 2 Tunneling Protocol over IPsec (L2TP/
   IPsec), 38
least privilege principle, 2, 17, 99, 168
   review of, 18
Legal, Regulations, Investigations, and
   Compliance domain, 217
legislative compliance of organization, 52
"lessons learned" review, 177, 210
licensing, 219–222
likelihood values in risk assessment, 70
limit checks, 101
link encryption, end-to-end encryption
   vs., 129
Linux, discretionary access controls, 5
Locard's exchange principle, 228
location-based authorization controls, 10
locks and keys, 254
logging, 11–12
   chain of custody, 230–231
   review of, 18
   system, 22
logical controls, 3
logical link control (LLC) sublayer, 28
logic bombs, 13, 101, 179
logo, legal protection for, 220
logs
   data capture from, 235
   Dynamic Host Configuration Protocol
      (DHCP), 236
   management, 11
   physical access control, 252
loss after countermeasure, 80
loss before countermeasure, 80

M

macros, 98
maintenance
   in development life cycle, 91
   of software, 93
malware, 13, 81, 101, 159, 178
   developers' use of bogus
      information, 237
   incident involving, 177
   types, 179
   USB-based, 35
   USB drives as risk, 173
management controls, 3
mandatory access controls (MACs), 5
man-in-the-middle attack, 43

maturity models, 92
maximum tolerable downtimes (MTDs), 196,
   200–201
maximum tolerable period of disruption
   (MTPOD), 200
MBSA tool, 184
media
   analysis in forensic procedures, 233–235
   management, 173
   protection based on data contents, 171
media access control (MAC) filtering, 33
media access control (MAC) sublayer, 28
medium, for steganography, 140
memory, application deallocation of, 153
memory leaks, preventing, 153
mesh network topology, 28
Message Analyzer (Microsoft), 236
message digest, 115–116
Message Digest 5 (MD-5), 96, 116
metrics, for security management, 81–82
MH227, 259
Microsoft
   Active Directory, 7, 119
   Active Directory Certificate Services (AD
      CS), 136
   BitLocker Drive Encryption, 106
   Group Policy, 187
   Identity and Access Management
      Series, 21
   Lync Server, 38
   Message Analyzer, 236
   Network Monitor, 236
   server products, 191
   TechNet, 34
   trial versions of software, 221
microwave transmitters, 252
mirrored sites, 205
mirroring, 188
   remote, 204
mission of organization
   security function and, 48
mobile code, 13, 98
mobile devices
   protection, 35
   use in public place, 262
mobile sites, 205
module test, 101
monitoring, 11–12
motion or movement detectors, 251
multifactor authentication, 6
   for VPNs, 38

Multilevel security mode, 148
multimedia collaboration, 37–38

N

National Institute of Standards and
   Technology (NIST), 5
   cryptographic standards, 108–109
   on block ciphers, 113
   on risk management, 49
   on system development life cycle, 91
   SP 800-30, on vulnerabilities, 68
   SP 800-34, on contingency
      planning, 197
   SP 800-34, on recovery procedures, 210
   SP 800-61, on incident response, 175
   SP 800-66, on HIPAA Security Rule, 240
   SP 800-88 on sanitization, 172
   SP 800-122, on PII, 52, 177, 262
natural access control, 248
natural disasters, 175
   risks, 248
natural surveillance, 248
natural territorial reinforcement, 248
natural threats, 68
need-to-know, 168
negligence, 55
Nessus, 102, 184
NetFlow (Cisco), 236
network access control devices, 33
network address translation (NAT), 30
network analysis, 235–236
network attacks, 41–44
network components, 31–36
   in server room, 256
network interface cards (NICs), 180
Network layer (OSI), 28
Network Monitor (Microsoft), 236
networks, secure architecture, 25–31
New Technology File System (NTFS), 106
Nmap, 15, 102
non-discretionary access controls, 5
non-repudiation, digital signature and, 121,
   123–124
NOOP sled, 99
normalization, 97
Northrup, Tony, 34

O

objectives of organization, security function
   and, 48

objects
   in Access Control policies, 3
   grouping, 5
   in security model, 146
Office Communications Server, 38
Office of Management and Budget
   (OMB), 61
one-way encryption, 117
online analysis processing (OLAP)
   databases, 160
Online Certificate Status Protocol
   (OCSP), 138
online transaction database (OLTP), 97
online transaction processing (OLTP)
   databases, 160
OpenPGP Alliance, 133
open ports, identifying, 102
Open Shortest Path First (OSPF), 32
open source licenses, 221
Open Systems Interconnection (OSI)
   model, 26
Open Web Application Security Project
   (OWASP), 96, 159
operating system
   patch management, 182–183
   security kernel in, 11, 156
operation of software, 93
operations, in development life cycle, 91
Operations Security domain, 167
Orange Book standard, 149
Organisation for Economic Co-operation and
   Development (OECD), 223
organization
   code of ethics, 227
   legislative and regulatory
      compliance, 52
   location, and exposure to outages, 201
   privacy requirements compliance, 52
   processes for security governance, 50
   restricted areas protection, 253
outages, metrics on, 81

P

packet analyzer, 235
packets, 28
paging file, 154
pan-tilt-zoom (PTZ) cameras, 251
parallel tests, 212
parental consent for website information
   collection, 225

passive infrared (IR) sensors, 250
passive response by IDS, 181
password crackers, 13
passwords, 6, 96
    masking, 256
    policies, 20
    reset systems for, 21
    storage, 126
patch management, 182–183
patch panel, 31
patents, 219, 220
    on software, 222
payload, for steganography, 140
Payment Card Industry Data Security Standard (PCI DSS), 61, 149, 151, 240
peer-to-peer (P2P) computing systems, 162
penetration tests, 84, 184
perimeter intrusion detection assessment systems (PIDAS), 252
perimeter security, 250–252
Permanent Virtual Circuit (PVC), 30
permission creep, 20
permissions, 169
personal identification number (PIN), 6
personal identity verification (PIV) cards, 253
personally identifiable information (PII), 52, 176, 224
personnel
    code of ethics and decisions of, 227
    fake phishing emails as test, 83
    in disaster recovery process, 208
        communication, 208–209
    notification in disaster recovery response, 207
    privacy and safety, 261
    termination processes, 76–77
    training in disaster recovery, 211
personnel security, 74–78
    education, training and awareness, 78–79
    employee termination processes, 76–77
    employment agreements and policies, 75–76
    screening of job candidates, 74–75
phishing emails
    fake as employee test, 83
    targeting executives with, 159
photoelectric fire detectors, 259
physical access control, logs, 252
physical addressing, 28
physical controls, 3

Physical (Environmental) Security domain, 247
physical intrusion detection systems, 250
Physical layer (OSI), 28
physical security, 31, 248–249
    controls, 250–252
    design for, 247–249
    implementation and operation, 255–260
    keys and locks, 254
    layers of, 249
    restricted and work area security, 256–257
PIN (personal identification number), 6
ping, 34
plain old telephone service (POTS), 36
plaintext data, 110, 111
planning, in development life cycle, 90
Platform as a Service (PaaS), 243
plenum-safe cable, 259
plenum space, 32
Point-to-Point Protocol (PPP), 38
Point to Point Tunneling Protocol (PPTP), 38
polyinstantiation, 161
Portable Network Graphics (png), 27
port mirroring, 31, 235
ports, 34
    closing unneeded, 42
    for TCP and UDP, 29
port scanners, 102
positive security model, 147–148
power line conditioners, 261
Presentation layer (OSI), 27
Pretty Good Privacy (PGP), 133–134
preventive controls, 4
preventive measures for attacks, 180–182
primary key in database, 97
privacy
    legal protection of data, 224–225
    of personnel, 261–263
Privacy Act of 2005, 52
privacy requirements compliance, 52
private branch exchange (PBX), 37
private key, 113–114
private ports, 29
privileged accounts, 21
privileges, monitoring, 169–170
probability values in risk assessment, 70
procurement processes, ensuring security in, 242
product evaluation models, 149–150

professional ethics, 225–227
programming languages, security issues, 98
projected benefits, 80
promiscuous mode, NICs in, 180
proprietary data, 64
Protection Profile (PP), 150
protocol analyzers, 14, 36, 180
protocol data unit (PDU), 27
protocols, implications of multilayer, 30
prototype test, 90
Prototyping model, 91
provisioning life cycle, 19–23
proximity badges, 252
proximity volumetric detectors, 251
proxy servers, 34
pseudorandom keys, 119
public information, use for hiring decisions, 75
public key, 113–114
    encrypting symmetric key with, 134
public key infrastructure (PKI), 40, 133, 135–137
public switched telephone network (PSTN), 36
purging media, 171

Q

qualitative risk assessment, 69, 70
Quality of Service (QoS) technologies, in ATM, 30
quantitative risk assessments, 69

R

radio frequency interference (RFI), 32
RAID (redundant array of independent disks), 187
RAID-0, 188
RAID-1, 188
RAID-3, 189
RAID-5, 189
RAID-10, 189–190, 190
rainbow tables, 6, 126
random access memory (RAM), 154
rapid application development (RAD), 91
RC4 encryption algorithm, 112
Real-time Transport Protocol (RTP), 37
reciprocal agreement between companies, 205
record retention policies, 172
recovery agent, 120

recovery controls, 4
recovery objectives, 201
recovery strategies, 196
redundant systems, 58
registered ports, 29
registered trademarks, 220
registration authority (RA), 135
regulations
    compliance requirements and procedures, 239–241
    audits, 240–241
    on reporting, 241
regulatory compliance of organization, 52
regulatory environment, 239–240
regulatory security policies, 60
relevant evidence, 231
reliable evidence, 231
remote access, 38–39
remote access protocols, 8
Remote Authentication Dial-in User Service (RADIUS), 8
remote journaling, 204
remote mirroring, 204
Remote Procedure Calls (RPCs), 97
remote wipe, 35
reports
    regulations on, 241
    security incidents, 176–177, 231–232
requirements analysis, in development life cycle, 90
reset systems for passwords, 21
residual risk, 71, 72
resources, protection, 173–174
restoration in disaster recovery, 209–210
restricted area security, 256–257
retinal scans, 7
return on investment (ROI) of control, 14
reverse engineering, 96
reviews, for testing BCP, 211
revocation of account access, 22
RFC (Request for Comments)
    1122, on TCP/IP, 26
    1123, on TCP/IP, 26
    1321, on MD5, 116
    1918, on private IP address ranges, 30
    2246, on TLS, 130
    2560, on OCSP, 138
    2743, on GSS-API, 95
    2744, on C language interfaces, 95
    2865, on RADIUS, 8
    2868, on RADIUS, 8

3575, on RADIUS, 8
3588, on Diameter, 8
3851, on S/MIME, 133
4301, on IPsec, 39
4880, on OpenPGP Alliance, 133
5080, on RADIUS, 8
5246, on TLS, 130
6101, on SSL, 130
6151, on MD5, 116
RFID database systems, for assets, 260
Rijndael algorithm, 112
ring network topology, 28
risk, 12, 48
    of data emanation, 157
    of goodwill loss, 176
    natural disaster, 248
    organization protection against, 55
    USB flash drives and, 173
risk assessments/analysis, 69–70
risk management, 67–74
    asset valuation, 14–15
    assignment/acceptance, 71–72
    countermeasure selection, 72–73
    threats and vulnerabilities identification, 68–69
    vulnerability analysis and, 15
risk mitigation, 71
risk transference, 71
rogue access point, 33, 256
role-based access control (RBAC), 5, 9, 20
root cause analysis, 177–178
root certificates, 135
routers, 31, 32
Routing Information Protocol 2 (RIP2), 32
RPO, 201
RSA (Rivest, Shamir, and Adleman) algorithm, 114–115
RSA Factoring Challenge, 115
RSA token, 6
RTO, 201

S

Safe Harbor framework, 223
safety of personnel, 261–263
salami attack, 218
salting hash, 6, 126
sandboxing, 98
sanitization, 171, 172
SANS Institute Reading Room, 16, 82
    computer incident response team overview, 229
    on electronic data retention policy, 172
    on physical security in data centers, 257
SANS Institute, security policy templates, 60
Sarbanes-Oxley Act of 2002 (SOX), 54, 67, 241
scope
    of project, development and documentation, 196–197
    of security incident, 176
screening of job candidates, 74–75
screen scraping, 39
SearchSecurity, 179
secret key cryptography, 113
Secure FTP (SFTP), 106
Secure Hashing Algorithm (SHA), 40
Secure Hashing Algorithm 1 (SHA1), 116
Secure Hashing Algorithm 256 (SHA-256), 96
Secure Multipurpose Internet Mail Extensions (S/MIME), 133
Secure Shell (SSH), 39, 106
Secure Sockets Layer (SSL), 38, 39, 107, 130–131, 158
Security Architecture & Design domain, 145
security architectures, vulnerabilities of, 155–158
Security Assertion Markup Language (SAML), 9, 158
Security Assurance Requirements, 150
security controls, 3–5
Security Functional Requirements, 150
security function, management of, 79–84
security governance, 49–56
    organizational processes, 50
security guards, 250
security incidents, 81. *See also* investigations of security incidents
    detection, 175
    determine the source of, 232
    handling and response, 230
    preventative measures, 178–182
    recovery, 177
    remediation and review, 177–178
    reporting, 176–177
    reporting and documenting, 231–232
    response management, 174–178
    scope of, 176
    verifying occurrence, 230

security information and event management
  (SIEM), 35
security information management (SIM), 35
security kernel, 11, 156
security management
  budget, 80–81
  metrics, 81–82
  resources, 82–83
security models, 145–149
security operations, 167–173
  completeness and effectiveness
    of, 83–84
  job rotation policy, 170
  monitoring privileges, 169–170
  record retention policies, 172
  roles and responsibilities, 51
  separation of duties and responsibilities
    policy, 169
security policies, 1–3, 49
  development and
    implementation, 59–63
  documentation, 62–63
Security Target (ST), 150
security triad, 56–58
seed value for stream cipher, 113
segment, 27
senior management
  organization security and, 59
  responsibility for information
    classifications, 64
  role in security, 51
sensitive information
  marking, handling, storing and
    destroying, 170
  storage of backups, 203
separation of duties policy, 2, 169
server-based vulnerabilities, 160
server rooms, physical security for, 255–256
servers, fault tolerance for, 190–191
service-oriented architecture (SOA), 157
Service Provisioning Markup Language
  (SPML), 9
service set identifier (SSID), disabling
  broadcast, 33
session key cryptography, 113
Session layer (OSI), 27
shareware, 221
shielded enclosures, 157
shielded twisted-pair (STP) cable, 32, 157
shoulder surfing, 126, 256
shredders, 171

shrinkwrap contract, 222
Shuart, Donna M., 43
side-channel attacks, 128
signature-based IDS, 180
signed applets, 159
Simple Integrity Axiom, 146
Simple Mail Transfer Protocol (SMTP), 27
Simple Network Management Protocol
  (SNMP), 29
Simple Security Rule, 5, 146
single loss expectancy (SLE), 72
single point of failure (SPOF), 195, 199
  eliminating, 157
single sign-on (SSO) authentication, 1, 7
site design, for security, 247–249
slack space, 233, 234
smart cards, 6, 253
  attacks on, 128
smartphones, retrieving information
  from, 238
smoke detectors, 259
smurf attack, 42
sniffers, 14, 180, 235
sniffing attacks, 36
SOAP, 97
social engineering, 13, 126, 256
  phone calls, 16
social media, search by potential
  employers, 75
software
  assessing security
    effectiveness, 100–102
  change management in
    development, 93–94
  copyright, 221
  licensing, 221
  security of environment, 95–97
  vulnerabilities and threats, 158–162
software analysis, 236–237
Software as a Service (SaaS), 243
software development life cycle
  (SDLC), 89–94, 90–91
Software Engineering Institute (Carnegie
  Mellon), 92
software keys, 174
source code
  obfuscation, 96
  review, 101
  security issues, 99
  signing, 98
spam, 160, 180
  metrics for, 81

spam over instant messaging (SPIM), 38
spear phishing, 159
spiral model, 91
spoofing, 13, 43–44
spyware, 179
Stacheldraht, 42
stack overflow error, 99
Standard Desktop Configuration (SDC), 61
standards, 60–61
star network topology, 28
stateful inspection, 34
steering committee, 50
steganography, 139–140
stream ciphers, 113
stripe of mirrors, 190
striping, 188, 189
striping with parity, 189
subject, of security model, 146
subjects
    grouping, 5
    in access control policies, 3
surge protectors, 261
swap file, 154, 237
Switched Virtual Circuit (SVC), 30
switches, 31
Symantec, 19, 133
symmetric cryptography, 111–113, 114
symmetric key, encrypting with public key, 134
synchronization of log times, 11
Synchronous Digital Hierarchy (SDH) fiber WANs, 30
Synchronous Optical Networking (SONET), 30
SYN flood attack, 14, 41
SYN (synchronize) packet, 29
Sysinternals, 237
System Center Configuration Manager (SCCM), 183
system development life cycle, 89, 90
system event managers (SEMs), 35
system high-security mode, 148
system images, 186
system logging, 22
system resilience, 187–191
system security architecture, 155–157
system security policies, 60
system vulnerabilities and threats, 158–162

T

tablet devices, retrieving information from, 238
tabletop exercises, 212
TACACS, 29
TACACS Plus (TACACS+), 8
tangible assets
    valuation, 73
tapes for backups, 203
    rotating out of service, 173
tape vaulting, 203
target of evaluation (TOE), 150
technical controls, 3
technical recovery teams, 208
technology, and process integration, 157
Telecommunications and Network Security domain, 25
teleconferencing, 37
temporal-based authorization controls, 10
termination process for employees, 76–77
territorial reinforcement, natural, 248
testing
    backup strategy effectiveness, 203
    black, white, or gray box, 184
    changes, 186
    in development life cycle, 90
thin clients, remote access use by, 39
third-party governance, 65–67
threats
    identifying, 68–69
    modeling, 13–14
    software and system, 158–162
threat-source, 69
three-way handshake, 29
time of check (TOC), 156
time of use (TOU), 156
time stamps, 95
    as evidence, 235
tort laws, 239
total risk, 71, 72
Towers of Hanoi tape rotation scheme, 203
tracert, 34
trademarks, 219, 220
trade secrets, 219, 220, 222
training
    in disaster recovery, 211
    of employees on security, 78–79
    for incident response team members, 229
    on laptop theft prevention, 260–261
    on safety and security issues, 262

transaction logs for databases, 97
trans-border data flow, 223–224
Transmission Control Protocol (TCP), 29
Transmission Control Protocol/Internet Protocol (TCP/IP) model, 26
transmission media, 32–33
Transport layer (OSI), 27
Transport Layer Security (TLS), 39, 107, 130–131, 158
transport mode in IPsec, 39
trial software, 221
Tribe Flood Network (TFN), 42
Trinoo, 42
Triple DES (3DES), 112
Trivial File Transfer Protocol (TFTP), 29
Trojan horse, 13, 179
TrueCrypt, 106
Trusted Computer System Evaluation Criteria (TCSEC), 149
Trusted Platform Module (TPM), 106, 155
tunneling, 156
tunneling protocols, 38
tunnel mode in IPsec, 39
twisted-pair cable, 32
two-factor authentication, 6, 7
TwoFish, 112
Type 1 errors, 7
Type 2 errors, 7

U

unauthorized access, protection against, 253
unauthorized accounts, 18
unhandled error, 99
uninterruptible power supply (UPS), 261
U.S. Air Force, Standard Desktop Configuration (SDC)
　Standard Desktop Configuration (SDC), 61
U.S. Department of Defense, document 5200.28, 148
U.S. Federal Bureau of Investigation (FBI)
　Intellectual Property Theft website, 221
　on safety for U.S. students traveling abroad, 262
U.S. Nuclear Regulatory Commission, on intrusion detection systems, 252
United States Privacy Act of 2005, 224
unregistered trademarks, 220
unshielded twisted-pair (UTP) cable, 32

updates to operating system and applications, 182
USB devices
　controls to restrict use, 35
　risk from, 173
user accounts
　de-provisioning, 21
　disabling during exit interview, 76
　managing changes, 17
　of ex-employees, 18
　review, 21–22
　revocation of access, 22
User Data Protocol (UDP), 27, 29
user entitlement, 17–18
user interface, constrained, 9
utilities, 257

V

vacation, mandatory policies on, 170
validation
　of certificates, 138–139
　of input, 99, 101
validity of data, 95–96
valuation of assets, 14–15
variables, names in code, 96
varifocal lenses, 251
VBScript, 98
vendors, 77
VeriSign Trust Network Certification Practice Statement document, 136, 137
version control for BCP, 213
video
　hiding data in, 139
　on security, 123
video conferencing, 37
views for databases, 147
virtualization, 154
virtual local area networks (VLANs), 40
virtual memory, 153
virtual private network (VPN), 38
virtual solutions for business continuity, 206
virtual team response, 228–229
viruses, 13, 179
visibility of people, facility design to maximize, 248
visitors, escort requirements for, 253–254
VM escape attack, 39, 154
VMware, 206
voice communications, 36–37
Voice over Internet Protocol (VoIP), 37

volatile memory, data stored in, 153
volumetric sensors, 251
vulnerabilities
    attack on unknown, 179
    of data stored in cloud, 162
    identifying, 68–69
    in IPv6, 156
    management, 183–184
    of security architectures, 155–158
    server-based, 160
    software and system, 158–162
vulnerability analysis, 15

W

warm sites, 205
Waterfall model, 91
water leaks or flooding, 258
watermarking, 140
web-based languages, vulnerabilities, 158
websites, vulnerabilities, 98
web spoofing attack, 43
well-known ports, 29
whaling, 14, 159
white box testing, 184
whitelisting, 148
Wi-Fi Protected Access 2 (WPA2), 33
window circuit, 250
Windows
    discretionary access controls, 5
    New Technology File System
        (NTFS), 106
windows, in building design, 248
Windows Live Messenger, 37
Windows Software Update Services
    (WSUS), 183
Wired Equivalent Privacy (WEP), 33, 113, 127
wireless access points (WAPs)
    isolation mode, 33
wireless access point (WAP), 31
wireless networks, 32–33
wireless routers, 32
Wireshark, 37, 236
wiring closet, best location for, 256
work area security, 256–257
worms, 13, 176, 179
write blocker, 234

X

X.509 standard, 137
XML exploitation attacks, 158
XML (Extensible Markup Language), 158
    standards based on, 9
XSS (cross-site script), 44, 98
    preventing attacks, 101

Z

zero-day exploit, 179
zero knowledge test, 184
Zimmermann, Phil, 133
zombies, 42, 43, 236

About the Author

DARRIL GIBSON, Security+, A+, Network+, CASP, SSCP, CISSP, MCT, CTT+, MCSE, MCITP, is founder and CEO of Security, Consulting, and Training, LLC. Darril has written or co-written more than 25 books, including several on security and security certifications. He regularly posts articles on *http://blogs.GetCertifiedGetAhead.com* and can be reached at *darril@GetCertifiedGetAhead.com*.

What do you think of this book?

We want to hear from you!
To participate in a brief online survey, please visit:

microsoft.com/learning/booksurvey

Tell us how well this book meets your needs—what works effectively, and what we can do better. Your feedback will help us continually improve our books and learning resources for you.

Thank you in advance for your input!

CPSIA information can be obtained at www.ICGtesting.com
Printed in the USA
LVOW101719140613

338664LV00018BA/618/P